Wenjin Ding

Simulation-Assisted Design of Polycrystalline Zeolite Catalysts

Simulation-Assisted Design of Polycrystalline Zeolite Catalysts

by
Wenjin Ding

KIT Scientific Publishing

Dissertation, Karlsruher Institut für Technologie (KIT)
Fakultät für Chemieingenieurwesen und Verfahrenstechnik (CIW), 2016

Tag der mündlichen Prüfung: 14. April 2016
Referenten: Prof. Dr.-Ing. Roland Dittmeyer,
Prof. Dr. Dr.h.c. Frerich Keil

Impressum

KIT Scientific Publishing

Karlsruher Institut für Technologie (KIT)
KIT Scientific Publishing
Straße am Forum 2
D-76131 Karlsruhe

www.ksp.kit.edu

Print on Demand 2016

ISBN 978-3-7315-0533-4
DOI 10.5445/KSP/1000055076

Simulation-Assisted Design of Polycrystalline Zeolite Catalysts

zur Erlangung des akademischen Grades eines

Doktors der Ingenieurwissenschaften (Dr.-Ing.)

von der Fakultät für Chemieingenieurwesen und Verfahrenstechnik
des Karlsruher Instituts für Technologie (KIT)

genehmigte

Dissertation

von

Wenjin Ding

aus Zhejiang, VR China

Tag der mündlichen Prüfung: 14. April 2016
Erster Gutachter: Prof. Dr.-Ing. Roland Dittmeyer
Zweiter Gutachter: Prof. Dr. Dr.h.c. Frerich Keil

Kurzfassung

Für die chemischen Industrie zeigen Zeolithmembranen vielversprechende Anwendungsmöglichkeiten in der Katalyse und für Trennprozessen. Im Gegensatz zu anderen mikroporösen Membranen wie z.B. Silikamembranen haben die Zeolithmembranen oft eine polykristalline Struktur mit koexistieren intrakristallinen Mikroporen und interkristallinen Mesoporen (d.h. Defekte, wie z.B. Risse). Wenn die Zeolithkristalle (z.B. ZSM-5) in der Zeolithmembran eine anisotrope Mikroporenstruktur aufweisen, sind die Diffusionskoeffizienten richtungsabhängig. Die polykristalline Natur der Zeolithmembranen und die anisotrope Struktur der Zeolithporensysteme hat einen signifikanten Einfluss auf deren katalytisches Verhalten und deren Trennleistungen durch die Beeinflussung des Massentransports im Porensystem.

In dieser Dissertation wurde eine simulationsgestützte Entwicklungsmethode basierend auf Experimenten und numerischen Simulationen entwickelt. Diese Methode unterstützt die Entwicklung von hierarchisch strukturierten Katalysatorsystemen auf der Basis polykristalliner und anisotroper Zeolithmembranen (oder Zeolithschichten), durch die Vorhersage einer optimalen Katalysatorstruktur und durch die Reduzierung der Trial-and-Error-Tests. Erfolgreich angewendet wurde diese Methode, sowohl bei der Optimierung von bifunktionellen Kern-Schale-Katalysatoren ($Cu/ZnO/Al_2O_3$@H-ZSM-5) mit einer Zeolith H-ZSM-5 Schaleschicht zur direkten Dimethylether (DME) Synthese aus Synthesegas, als auch bei der Analyse der Leistungsfähigkeit einer katalytischen H-ZSM-5-Membran für die Xylolisomerisierung.

Ein Eckpfeiler dieser Dissertation ist ein neues dreidimensionales Porennetzwerkmodell, mit dem man die Diffusion und die chemische Reaktion in polykristallinen und anisotropen Zeolithmembranen (oder Zeolithschichten) simulieren kann. Der diffusive Transport in den intrakristallinen Mikroporen wird mit dem verallgemeinerten Maxwell-Stefan Oberflächendiffusionsmodell (surface diffusion model, SDM) oder dem Single-File-Diffusions-Modell (single-file diffusion model, SFDM) modelliert, während das Knudsen-Diffusion-Modell (KDM) zur Berechnung der Diffusionsvorgänge in den interkristallinen Mesoporen zur Anwendung kommt. Das oben beschriebene Modell ist für viele praxisbezogene Probleme anwendbar, da chemische Reaktionen, die durch unterschiedliche kinetische Modelle beschrieben werden, implementiert werden können. Zusätzlich ermöglicht es die polykristallinen Zeolithe mit einem 3-D Kristallit-Porennetzwerk-Modell (CPNM) abzubilden. Im Vergleich zu früheren Porennetzmodellen, berücksichtigt das CPNM den anisotropen Transport innerhalb der Zeolithkristalle dadurch, dass man jedem Kristall zwei richtungsbeschreibende Parameter zuordnet. Zusätzlich kann man sowohl den Einfluss der Zeolithkristallgröße und der Konnektivität der interkristallinen Mesoporen (d.h. der Defekte) mit den intrakristallinen Mikroporen an der Leistungsfähigkeit der polykristallinen Zeolithmembranen als auch die hierarchisch strukturierten Katalysatorsysteme mit den Zeolithschichten untersuchen.

Das SDM und SFDM geht davon aus, dass die Diffusion der Moleküle im adsorbierten Zustand an der Porenwand stattfindet. Aus diesem Grunde hängen die Diffusionsflüsse von der Menge der adsorbierten Moleküle und ihrer Beweglichkeit in adsorbiertem Zustand ab. Somit sind zuverlässige Adsorptions- und Diffusionsparameter für die Simulation mit beiden oben genannten Modelle erforderlich. Aus diesem Grunde wurde eine Hochtemperatur-Hochdruck-Adsorptionsvorrichtung basierend auf einer Hochfrequenzschwingmikrowaage (Langatate Kristall-Mikrowaage, LCM) aufgebaut, um die Adsorptionsparameter von Gasen in H-ZSM-5 im der Nähe der Reaktionstemperaturen (z.B., 200-300 °C zur direkten DME-Synthese) experimentell zu ermitteln. Damit wurden die Adsorptionsisothermen der reinen Komponenten von CO_2, H_2O, Methanol und DME in H-ZSM-5 im Temperaturbereich von 50 bis 150 °C und bei Drücken bis 18 bar messtechnisch bestimmt. Für Permeationsmessungen durch eine H-ZSM-5-Membran ist eine Hochtemperatur-Hochdruck-Wicke-Kallenbach-Zelle (WK-Zelle) aufgebaut worden. Durch Auswertung der Messergebnisse konnten, die für das SDM und das SFDM benötigten korrigierten Oberflächendiffusionskoeffizienten, ermittelt werden.

Für die Simulation der direkten DME-Synthese in den Kern-Schale-Katalysatoren auf Basis der CPNM sind für die Optimierung zuverlässige kinetische Parameter der chemischen Reaktionen, die in den Katalysatoren auftreten, erforderlich. Die Berechnung mit dem CPNM ist sehr zeitaufwendig, insbesondere wenn große Netzwerke des CPNM für die Simulation gewünscht sind. Um die Rechenzeit der Simulationsmodelle auf der Basis des CPNM zu reduzieren, kann man die Parallel-Computing-Technologie in Verbindung mit Multicore-Super-Computer verwenden. Da die Anwendung des CPNM zur Simulation der Diffusion und der chemischen Reaktion in dem Kern-Schale-Katalysator numerisch zu aufwendig ist, wurde im ersten Schritt ein einfacheres heterogenes Modell, unter der Annahme einer pseudohomogenen Schale, entwickelt. Dieses soll zur Bestimmung der Reaktionsparameter in dem Kern-Schale-Katalysator verwendet werden. Wobei ausgehend von dem heterogenen Modell für einen mit dem Kern-Schale-Katalysator gefüllten Rohrreaktor ein Reaktormodell entwickelt wurde. Gegenüber dem Reaktormodell basierend auf dem CPNM (Berechnungszeit von über 40 Stunden), erzielt man mit der DME-Synthese im Rohrreaktor, gepackt mit den Kern-Schale-Katalysatorpartikel, Rechenzeiten von ca. 5 Minuten mit einem herkömmlichen Computer mit zwei CPU-Kernen. Das Simulationsmodell wurde verwendet, um die Reaktionsparameter für die Kern-Schale-Katalysatorpartikel zu bestimmen, die durch den Kooperationspartner - Institut für Chemische Reaktionstechnik (CRT) an der Friedrich-Alexander-Universität Erlangen-Nürnberg (FAU) synthetisiert wurden.

Das CPNM wurde angewendet, um sowohl die direkte DME-Synthese in bifunktionellen Kern-Schale-Katalysatorpartikel ($Cu/ZnO/Al_2O_3$@H-ZSM-5) als auch die Isomerisierung von Xylol in einer Katalytischen H-ZSM-5-Membran, angeordnet in einer WK-Zelle, zu simulieren. Die Wirkungen der polykristallinen und anisotropen Struktur von H-ZSM-5-Katalysatoren (z.B. Kristallorientierung) auf diese beiden Reaktionssysteme wurden mittels Simulationsrechnungen untersucht. Die Adsorptions-, Diffusions- und Reaktionsparameter entnahm man aus Experimenten oder der Literatur. Richtlinien zur Katalysatorherstellung und Optimierung konnten aus den Simulationsergebnissen abgeleitet werden.

Parts of this dissertation have already been published in

W. Ding, H. Li, P. Pfeifer, R. Dittmeyer, Crystallite-Pore Network Model of Transport and Reaction of Multicomponent Gas Mixtures in Polycrystalline Microporous Media, *Chem. Eng. J.*, **2014**, 254, 545-558.

W. Ding, M. Klumpp, S. Lee, S. Reuß, S. A. Al-Thabaiti, P. Pfeifer, W. Schwieger, R. Dittmeyer, Simulation of One-Stage Dimethyl Ether Synthesis over Core/Shell Catalyst in Tube Reactor, *Chemie Ingenieur Technik*, **2015**, 87, 702-712.

W. Ding, M. Klumpp, H. Li, U. Schygulla, P. Pfeifer, W. Schwieger, K. Haas-Santo, R. Dittmeyer, Investigation of High-Temperature and High-Pressure Gas Adsorption in Zeolite H-ZSM-5 *via* the Langatate Crystal Microbalance: CO_2, H_2O, Methanol, and Dimethyl Ether, *J. Phys. Chem. C*, **2015**, 119(41), 23478-23485.

W. Ding, G. Baracchini, M. Klumpp, W. Schwieger, R. Dittmeyer, Adsorption Device Based on a Langatate Crystal Microbalance for High Temperature High Pressure Gas Adsorption in Zeolite H-ZSM-5. *J. Vis. Exp.*, **2016**, 114, e54413, doi:10.3791/54413.

Abstract

In chemical industry, zeolite membranes have shown promising applications in catalytic and separation processes. Different from other microporous membranes like silica membranes, zeolite membranes often have a polycrystalline structure with coexisting intracrystalline micropores and intercrystalline mesopores (i.e., defects such as pinholes, cracks). Moreover, when the zeolite crystals in the zeolite membranes have an anisotropic micropore structure, e.g., ZSM-5, the diffusivities in the zeolite systems depend on the direction. Polycrystalline nature of the zeolite membranes and anisotropic structure of the zeolite pore systems are considered to have significant effects on the catalytic and separation performance via influencing the mass transport of molecules in the pore systems of the zeolite membranes.

In this dissertation, a simulation-assisted design method based on experiments and simulations was established to guide the development of hierarchically structured catalyst systems based on polycrystalline, anisotropic zeolite membranes (or zeolite layers) by predicting the optimal catalyst structure and reducing the trial-and-error testing. It has been used to assist the optimization of bifunctional core-shell catalysts ($Cu/ZnO/Al_2O_3$@H-ZSM-5) with a zeolite H-ZSM-5 shell layer for direct dimethyl ether (DME) synthesis from synthesis gas, as well as to analyze the performance of catalytic H-ZSM-5 membranes for xylene isomerization.

A cornerstone of this dissertation is a new three-dimensional (3-D) pore network model which was developed to simulate anisotropic multicomponent diffusion and reaction in polycrystalline zeolite membranes or layers. Transport in the intracrystalline micropores was modeled with the generalized Maxwell-Stefan surface diffusion model (SDM) or single-file diffusion model (SFDM), while the Knudsen diffusion model (KDM) was used to simulate the diffusion in the intercrystalline mesopores. This model is applicable to many practical problems since reactions represented by any type of kinetic expression are allowed. Moreover, it has a new feature that it represents the polycrystalline zeolites with a 3-D crystallite-pore network model (CPNM). In contrast to previous pore network models, the CPNM has the novel aspect of modeling the anisotropic transport inside the zeolite crystallites forming the polycrystalline zeolite membranes or layers by assigning to them two parameters to describe the orientation. In addition, it can investigate the effect of the size of zeolite crystallites and the connectivity of the intercrystalline mesopores (i.e., defects) with the intracrystalline micropores on the performance of the polycrystalline zeolite membranes, as well as the hierarchically structured catalyst systems based the zeolite layers.

Since the SDM and SFDM for diffusion in micropores consider that the molecules diffuse in adsorbed state on the pore wall, the diffusional fluxes of the molecules in the micropores therefore depend on the amount of adsorbed molecules and their mobility in adsorbed state. Thus, both reliable adsorption and diffusion parameters are required in the SDM and SFDM. In this dissertation, a high-temperature high-pressure adsorption device based on a high frequency oscillating microbalance (langatate crystal microbalance, LCM) was set up to determine the adsorption parameters of gases in H-ZSM-5 close to the reaction temperatures (e.g., 200-300 °C for direct DME synthesis). It has been used to measure the single component adsorption isotherms of CO_2, H_2O, methanol, and DME in H-ZSM-5 at temperatures between 50 and 150 °C and pressures up to 18 bar. Moreover, a high-temperature high-pressure Wicke-Kallenbach cell (WK) was established to measure the permeation of gases through an H-ZSM-5 membrane for determination of the corrected surface diffusion coefficients in the SDM and SFDM.

In the simulations of the direct DME synthesis in the core-shell catalysts based on the CPNM for the catalyst optimization, the reliable kinetic parameters of the reactions occurring in the catalysts are required. However, the calculation of the CPNM is very time-consuming, in particular when a large network size of the CPNM is required for the simulations. In order to reduce the calculation time of the simulation models based on the CPNM, parallel computing technology based on a multicore super computer is required. Here now, prior to the CPNM, a heterogeneous model under the assumption of a pseudo-homogeneous shell was used to simulate diffusion and reaction within the core-shell catalyst particles. For determination of the reaction parameters in the core-shell catalyst, a reactor model based on this heterogeneous model was developed for a tubular reactor, which was packed with the core-shell catalyst particles to obtain the catalyst test data. Compared to the reactor model based on the CPNM (calculation time of over 40 hours), it can simulate the DME synthesis in the tubular reactor packed with the core-shell catalyst particles within 5 minutes in a conventional personal computer only with two CPU cores. This simulation model has been used to determine the reaction parameters for the core-shell catalyst particles prepared by the cooperation partner – Institute of Chemical Reaction Technology (CRT) at Friedrich-Alexander-University Erlangen-Nürnberg (FAU).

The CPNM has been applied to simulate the direct DME synthesis in bifunctional core-shell catalyst ($Cu/ZnO/Al_2O_3$@H-ZSM-5) particles, as well as xylene isomerization in a catalytic H-ZSM-5 membrane placed in a WK cell. The effects of the polycrystalline and anisotropic structure of H-ZSM-5 catalysts (e.g., crystallite orientation) on these two reaction systems were investigated via simulations with the adsorption, diffusion, and reaction parameters obtained experimentally or from literature. Some guidances on the catalyst preparation and optimization were also given from the simulation results.

Contents

Abbreviations

ACS	American Chemical Society
AGIR	Combined IR & gravimetric analysis
BET	Brunauer-Emmett-Teller
CFD	Computational fluid dynamics
CPNM	Crystallite-pore network model
CRT	Institute of Chemical Reaction Engineering
CSTR	Continuous stirred-tank reactor
DFT	Density functional theory
DGM	Dusty-gas model
DME	Dimethyl ether
FID	Flame ionization detector
FT	Fischer-Tropsch
GC	Gas chromatography
GCP	Gas chromatography pulse
GDL	Gas diffusion layer
GHSV	Gas hourly space velocity
HPT	High pressure transducer

IAST	Ideal adsorbed solution theory
IKFT	Institute of catalysis research and technology
IMVT	Institute for Micro Process Engineering
KDM	Knudsen diffusion model
KIT	Karlsruhe Institute of Technology
LCM	Langatate crystal microbalance
LPG	Liquefied petroleum gas
LPT	Low pressure transducer
MFC	Mass flow controller
PFG-NMR	Nuclear magnetic resonance pulsed field gradient spectroscopy
PNM	Pore network models
QCM	Quartz crystal microbalance
QENS	Quasielastic neutron scattering
SAC	Steam-assisted crystallization
SDM	Surface diffusion model
SEM	Scanning electron microscope
SFD	Single-file diffusion
SFDM	Single-file diffusion model
STP	Standard temperature and pressure

TCD	Thermal conductivity detector
TPD	Temperature-programmed desorption
WDX	Wavelength-dispersive X-ray spectroscopy
WK-cell	Wicke-Kallenbach cell
XRD	X-ray diffraction
TEOM	Tapered element oscillating microbalance
ZLC	zero-length-column

Acknowledgments

First and foremost I would like to express my special appreciation and thanks to my supervisor Professor Dr.-Ing. Roland Dittmeyer. You have been a tremendous mentor for me. I would like to thank you for encouraging my research and for allowing me to grow as a research scientist. Your advice both on research as well as on my career have been priceless. You have taught me, both consciously and unconsciously, how to be a good research scientist. I would also like to give a special thanks to Professor Dr.-Ing. Peter Pfeifer for your advising and support in my experimental and simulation work. Without you, the experiments and my research cannot go so smoothly. Another special thanks, I would like to give to Professor Dr. Frerich Keil at Hamburg University of Technology, thank you for your time and work to be my second referee of my dissertation.

I thank very much my colleagues at IMVT, Mr. Seungcheol Lee, Dr. Katja Haas-Santo, Dr. Hui Li, Mr. Ulrich Schygulla, Mr. Messerschmidt, etc., as well as my cooperation partners at CRT at University Erlangen-Nürnberg, Professor Dr. Wilhelm Schwieger, Mr. Michael Klumpp, for the kind support and cooperation in my PhD research. I really enjoy the PhD study and research life with you.

The last thanks I would like to give to my wife, my daughter, my parents, my parents in-law: 首先，我要感谢我的爱人，孙佳妮女士，没有你的支持和鼓励，我不可能如此顺利地完成我的博士研究工作，并且获得如此多的收获，谢谢你为我所作出的辛苦付出！我还要感谢我心爱的女儿，你是我一往直前的力量源泉，也是我博士研究阶段中的又一个美好收获！最后，我要感谢我的父母和岳父母，感谢你们的支持，在我们最艰难的时候，帮助照顾我们心爱的女儿，让我们没有后顾之忧。我要感谢我的父母和奶奶，你们的养育之恩，还有给予我支持的亲人们。是你们对我一如既往的支持，才有了我的今天。

Karlsruhe, August 2016 Wenjin Ding

1 Introduction

1.1 Simulation-assisted design of porous heterogeneous catalysts

In recent years, the interest in simulation-assisted design of porous heterogeneous catalysts has grown rapidly with the increasing calculation speed and memory of computers, since it can accelerate the development of new catalysts or assist the improvement of existing catalysts by predicting the optimal composition, surface structure or pore structure of catalysts and eliminating or at least reducing the trial-and-error testing *via* modeling and simulations.

In the past few decades, one of the key developments in the simulation-assisted design of heterogeneous catalysts was the application of the density functional theory (DFT) in modeling of surface processes on catalysts [1]-[4], i.e., molecular adsorption/desorption, reaction, and diffusion on the catalyst surface. DFT is a computational quantum mechanical modeling method to investigate the electronic structure of the catalyst surface [3]. Thus, the DFT calculation can provide a molecular-level understanding of heterogeneous catalysis [3]-[5] on the catalyst surface. As an example for its application in the simulation-assisted design of porous heterogeneous catalysts, DFT has been used in the computational screening of new methanation catalysts to find the best catalyst with optimal metal composition, which performs the highest surface catalytic activity [4].

However, besides the surface catalytic activity, the pore structure of heterogeneous catalysts, particularly microporous zeolite catalysts [6]-[7] with pore sizes close to the molecular size, often have a significant effect on their catalytic performance, since their pore structures (e.g., pore size) affect the transport processes of molecules of reactants and products in the catalysts, significantly. It is therefore possible that one catalyst with a high surface catalytic activity however shows a poor catalytic performance due to the transport limitation in its poor pore structure. Thus, the simulation-assisted design of porous heterogeneous catalysts does not only require the understanding of the surface processes on the catalysts, but also of the pore structure of the catalysts and of the transport in the complex pore space *via* modeling and simulation.

1.2 Polycrystalline zeolite catalysts

In the chemical industry, zeolite materials have shown promising applications in catalytic and separation processes. In contrast to other porous heterogeneous catalysts like metal oxide catalysts, they generally have a microporous and polycrystalline structure, which is considered to have a significant effect on the catalytic performance due to the low molecular diffusivity and molecular-sieving effect in the micropores. In the following subsection, two catalyst systems with zeolite catalysts will be presented, which are thought to have promising applications in industry.

1.2.1 Bifunctional core-shell catalysts with a zeolite shell

It was reported in the literature [8]-[10] that compared to conventional admixed hybrid catalysts, bifunctional core-shell catalysts with a polycrystalline zeolite shell (see Figure 1-1) reach higher selectivity to the products for a consecutive reaction system, since all the intermediates produced in the core have to diffuse through the shell and simultaneously react to the products. For instance, Bao, et al. (2007) [8] coated an H-beta zeolite membrane directly onto the surface of Co/Al_2O_3 catalyst pellets to form a core-shell catalyst for direct synthesis of isoparaffins from synthesis gas based on Fischer-Tropsch (FT) synthesis. The prepared catalyst showed excellent performance such that the formation of C_{12+} hydrocarbons was suppressed completely and the middle isoparaffins became the main products. In 2010, Yang, et al. [9] reported that a bifunctional core-shell catalyst

(Core: $Cu/ZnO/Al_2O_3$ for methanol synthesis; Shell: zeolite ZSM-5 for methanol dehydration) for direct synthesis of dimethyl ether (DME) from synthesis gas had a strikingly higher selectivity to DME than the admixed hybrid catalyst.

Since these core-shell catalysts have the dominant transport resistance in the microporous zeolite shell (generally with few defect pores), the pore structure of the shell (e.g., size and concentration of defect pores, size and orientation of zeolite crystallites) and their thickness have a significant effect on the catalyst performance *via* affecting the transport of reactants and products through the shell. Thus, a realistic model, which can simulate the diffusion and reaction in the complex pore structure of the polycrystalline shells, is required in the simulation-assisted design of such core-shell catalysts. Moreover, reliable adsorption, diffusion, and reaction parameters are needed to account for the diffusion and reaction in the catalysts for catalyst optimization.

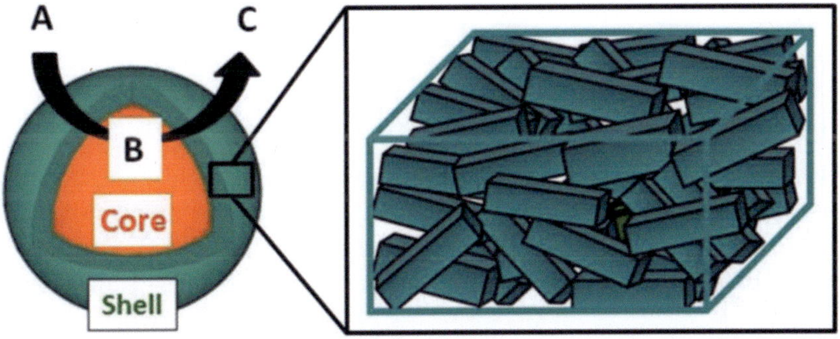

Figure 1-1: Representation of a bifunctional core-shell catalyst (left), which has a mesoporous core and a polycrystalline zeolite shell (right). For the direct synthesis of branched hydrocarbons from synthesis gas [8], core catalyst: Co/Al_2O_3 catalyst for Fischer-Tropsch synthesis; A: synthesis gas; B: linear hydrocarbons; C: branched hydrocarbons. For the direct synthesis of DME [9], core catalyst: $Cu/ZnO/Al_2O_3$; A: synthesis gas; B: methanol; C: DME.

1.2.2 Catalytic zeolite membranes

As mentioned above, zeolite materials have shown promising applications in separation and selective catalytic processes due to their molecular-sieving property and excellent thermal/chemical stability. Figure 1-2 shows the application of zeolite membranes as selective catalytic membranes in a catalytic membrane reactor. Owing to the separation of the products continuously by the zeolite membranes, thermodynamic limitations of the reaction on the conversion of reactants and the selectivity to the products can be overcome. For instance, it has been reported in [11]-[13] that high selectivity to the most industrially sought after *p*-xylene could be achieved by xylene isomerization in catalytic membrane reactors with ZSM-5 membranes, since *p*-xylene (d = 0.58 nm) has a smaller molecular size and therefore higher diffusivity in the ZSM-5 membranes than the other larger isomers, i.e., *o*- and *m*-xylene (both d = 0.68 nm).

The ZSM-5 membranes have not only a polycrystalline nature, but also an anisotropic structure, since the ZSM-5 crystallites forming the membranes have a three-dimensional anisotropic microstructure. As shown in Figure 1-3, the micropores in *b*-direction are straight, while those in *a*-direction are zig-zag. The effect of the anisotropic structure on the transport has been shown by Hong, et al. (1991) [14] and Kärger (1991) [15] *via* experiments and simulations. They found that the diffusivity of molecules (e.g., xylene, methane) in the straight micropores in *b*-direction is much higher than that in the zig-zag micropores in *a*-direction. Based on the anisotropic transport property of the ZSM-5 membranes, Lai, et al. (2003) [16] have prepared a *b*-oriented ZSM-5 membrane for separation of gas mixtures considering the higher diffusivity in the *b*-direction. The prepared membrane showed better permeation and separation performance for xylene isomers than randomly oriented membranes. Therefore, the simulation model used in the zeolite catalyst design should be able to simulate not only the polycrystalline structure but also the anisotropic structure of the zeolite membranes or layers.

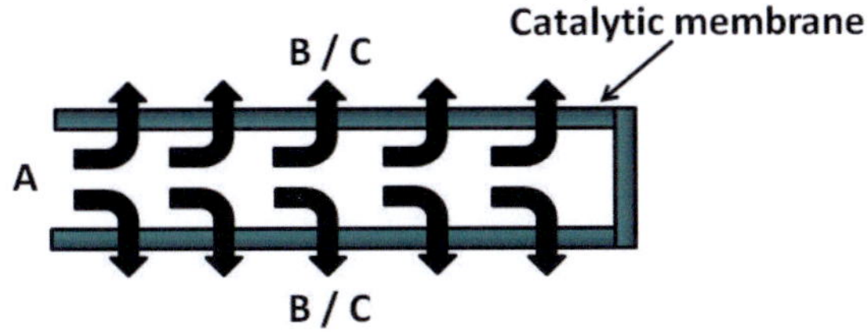

Figure 1-2: Representation of a catalytic membrane reactor with a zeolite membrane. For the selective xylene isomerization, A: *m*-xylene (or *o*-xylene); B: *o*-xylene (or *m*-xylene); C: *p*-xylene.

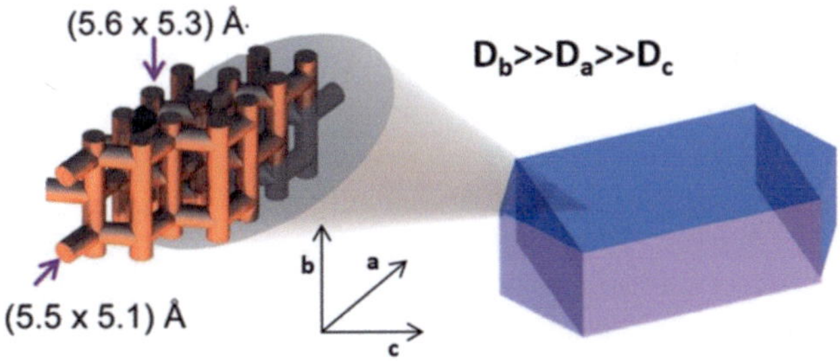

Figure 1-3: Three-dimensional anisotropic microstructure of a ZSM-5 crystallite (adopted from the dissertation of Koriabkina [17]), which has zig-zag and straight micropores in *a*- and *b*-direction, respectively.

1.3 Simulation-assisted design of zeolite catalysts – objectives and tasks

The objectives of the PhD study presented in this dissertation were to develop a catalyst design method based on simulations and experiments for zeolite catalysts with polycrystalline and anisotropic structure, and to apply this method to design the H-ZSM-5 shell in the core-shell catalysts ($Cu/ZnO/Al_2O_3$@H-ZSM-5) for the one-stage DME synthesis from synthesis gas.

Figure 1-4 shows an overview of this PhD study including the main tasks/objectives:

- Development of simulation models to account for diffusion and reaction in zeolite catalysts with polycrystalline and anisotropic structure,
- Determination of the required adsorption, diffusion and reaction parameters for the simulations, both experimentally and from literature,
- Validation of the developed simulation models with experimental data from the catalytic tests on the prepared catalysts, and
- Simulations using the developed simulation models and parameters obtained from experiments or literature to assist the optimization of zeolite catalysts.

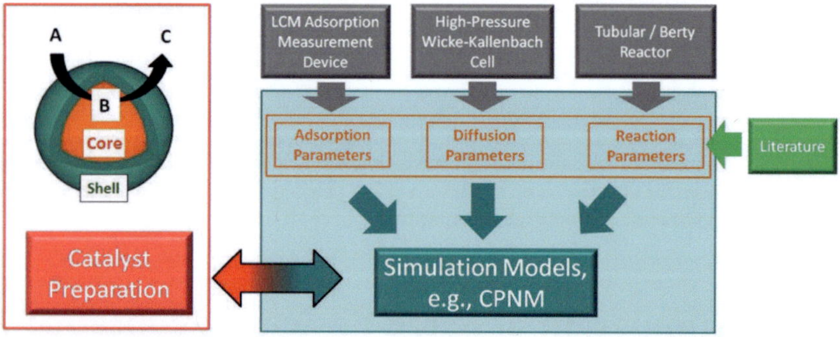

Figure 1-4: Overview of this PhD study for simulation-assisted design of catalyst systems with polycrystalline zeolite catalysts.

This PhD study was funded by German Research Foundation (DFG) under the program SPP 1570: Porous media with defined porous structure in chemical engineering - modeling, applications, synthesis (Di 696/9-1 and Di 696/9-2). The project was jointly executed with the Institute of Chemical Reaction Engineering (CRT) at University Erlangen-Nürnberg, who were in charge of the synthesis of the microsized core-shell catalysts for DME synthesis and the H-ZSM-5 membranes for the permeation tests in the WK cell.

The main achievement of the modeling work is the development of a more realistic pore network model – crystallite-pore network model (CPNM), which can simulate the polycrystalline and anisotropic structure of the ZSM-5 zeolite catalysts. Combining the kinetic and diffusion models, it could predict the effects of the structural parameters of the ZSM-5 catalysts, such as concentration of defects (intercrystalline pores), crystallite size and orientation, thickness of layers, on the catalytic performance.

For determination of adsorption and diffusion parameters for the simulations, a high-temperature high-pressure adsorption device based on a high frequency oscillating microbalance (langatate crystal microbalance, LCM) was devised together with colleagues and used to determine the adsorption parameters of related gases in the direct DME synthesis (e.g., methanol, DME) in H-ZSM-5 close to the reaction temperature. Moreover, a high-pressure and high-temperature Wicke-Kallenbach cell was established to measure the permeation of gases through the H-ZSM-5 membrane for determination of the corrected surface diffusion coefficients. These adsorption and diffusion parameters are required in the modeling of diffusion in the micropores of H-ZSM-5 by the generalized Maxwell-Stefan surface diffusion model (SDM) proposed by Krishna [18]. The SDM will be introduced in detail in Chapter 2.

In this dissertation, reactor models based on the CPNM and a heterogeneous model for the core-shell catalyst particles were established to simulate the direct DME synthesis in a tubular or Berty-type reactor placed with the the prepared core-shell catalyst particles. *Via* fitting with the experimental data from the tubular and Berty-type reactor, these reactor models were expected to be used to determine reaction parameters for the catalyst optimization *via* simulations. However, the calculation of the reactor models based on the CPNM is very time-consuming, in particular when a large network size of the CPNM is required for the simulations, and the concentration outside the catalyst particles varies in the reactors, e.g., the tubular reactor. Parallel computing technology and a multicore super computer is required to reduce significantly the calculation time of the reactor models based on the CPNM so that they can be used for the determination of reaction parameters. Therefore, the reactor model based on the heterogeneous model was used in this dissertation to determine the reaction parameters for the core-shell catalyst, which were prepared by CRT and investigated in the tubular reactor, since compared to the reactor model based on the CPNM (calculation time of over 40 hours), it can simulate the DME synthesis in the tubular reactor within 5 minutes using a conventional personal computer only with two CPU cores.

In this dissertation, *Chapter 2* reviews pore network models for porous materials and gas diffusion models in pores with various size available in literature.

The modeling work is introduced in detail in *Chapter 3* – a 3-D crystallite-pore network model for the polycrystalline and anisotropic zeolites.

Chapter 4 describes the experimental work carried out in the PhD study, i.e., the adsorption measurement device based on the LCM and the WK permeation cell for the determination of adsorption and diffusion parameters.

Chapters 5 shows the 1-D reactor model based on the heterogeneous model, which has been used to determine kinetic parameters of reactions in the prepared microsized core-shell catalyst particles for DME synthesis.

Chapters 6 presents the simulation results of the 3-D crystallite-pore network model for the core-shell catalysts for DME synthesis.

Chapter 7 shows the application of the crystallite-pore network model for xylene isomerization in a catalytic H-ZSM-5 membrane.

Finally, *Chapter 8* gives the conclusions and outlook of this PhD study.

2 Modeling of gas transport in porous media

In the preparation of porous catalysts, small pores are required in order to achieve a high surface area per volume, which leads to a high concentration of active sites, i.e., a high catalytic activity. However, such small pores often limit the supply of reactants and the removal of products from the active sites, which are typically dispersed on the surface of inner pores. Therefore, the pore structure of porous catalysts and transport in the pores often play an important role in heterogeneous catalysis in porous catalysts. In the simulation-assisted catalyst design, realistic models for the pore structure of porous catalysts and reliable models for transport in the pores are required to understand the influence of the pore structure on the catalyst performance. For modeling of gas transport in the complex pore space of porous media, various pore network models [19]-[32] and diffusion models ([18], [33]-[34], [36]-[37], [44], [49]) have been developed in the past few decades. In 2011, a nice review about modeling of diffusion and reactions in porous media was published by Keil [51]. In this chapter, the available pore network and diffusion models in the literature are reviewed.

2.1 Pore network models for porous media

2.1.1 Parallel pore network models

Pore network models (PNM) have proven to be a powerful tool to study the effect of the pore structures of porous media on the separation and catalytic performance [19]-[32]. Wheeler was the one of the first to have used a pore model to simulate diffusion and reaction in the porous catalysts. He used a parallel pore model as shown in Figure 2-1 for porous catalysts in 1951 [19]. In this model, parallel pores with the mean radius and length were used to simulate the pore space of the porous catalysts. The mean radius and length of the parallel pores were determined according to the assumption that the sum of the surface areas of all the pores constituting the pore network is equal to the Brunauer–Emmett–Teller (BET) surface area, and that the total pore volume is equal to the experimental pore volume. In this model, the pores are also allowed to have a radius with a distribution, e.g., a normal distribution.

Due to the calculation speed limitation of computers at that time, the pores in the parallel pore network model were assumed to be not interconnected for simplicity. Thus, this simple model cannot simulate the complex pore structure of real porous catalysts, e.g., it cannot investigate the effect of pore interconnectivity and dead ends. However, it was shown by pore network modeling and the percolation theory [20]-[25] that the pore interconnectivity and dead ends in the porous media have a significant effect on transport in the porous media.

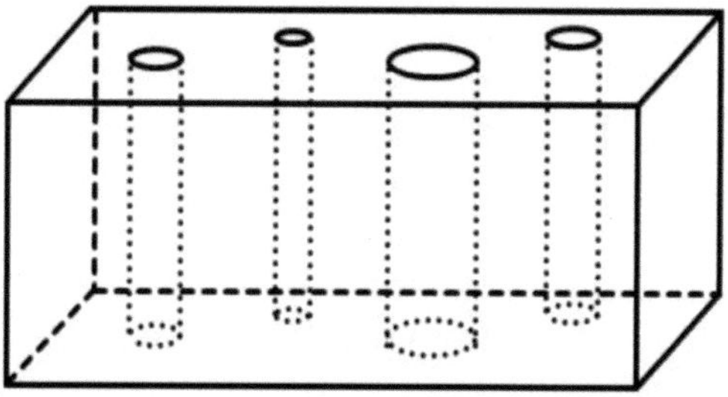

Figure 2-1: Parallel pore model for porous media.

2.1.2 Bethe pore network models

In 1989, Reyes, et al. [20] used a Bethe lattice to simulate the pore space of highly disordered porous catalysts. As shown in Figure 2-2, the Bethe network is an infinite branching tree without reconnecting branches and entirely characterized by a coordination number that describes the number of bonds emanating from each branching site (i.e., pore interconnectivity). In that work, it was suggested [20] that the pore radius can be determined from mercury porosimetry, while the pore interconnectivity could be either treated as an adjustable parameter or estimated from mercury porosimetry data or from percolation thresholds obtained in fragmentation studies.

Compared to the simple parallel pore model, the more complex Bethe PNM can describe the pore interconnectivity and account for the influence of tortuous trajectories and dead ends on transport rates. However, this network has a drawback that the infinite branches are not reconnected. This does not agree with the pore structure of real porous catalysts, which generally have highly interconnected pores.

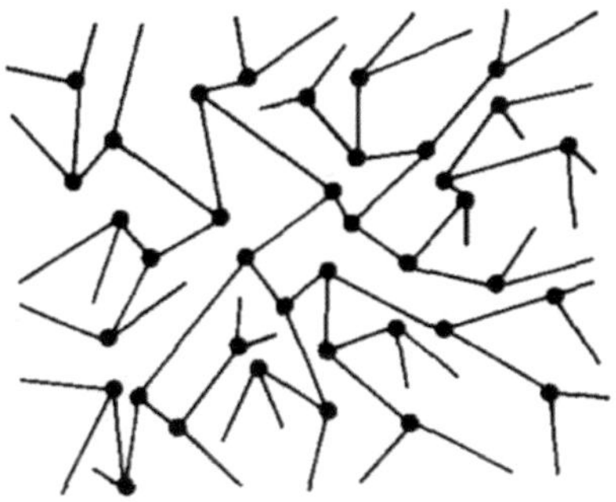

Figure 2-2: Bethe network with a coordination number of 3, adopted from [20].

2.1.3 Irregular random pore network models

In general, porous media like porous catalysts have highly interconnected pore throats and bodies as shown in Figure 2-3 left. Therefore, based on the concept that every porous medium can be mapped with an equivalent network of sites (pore bodies) and bonds (pore throats), Hollewand and Gladden (1992) [21] used an irregular random pore network model to model the pore space of the porous medium in Figure 2-3 left. The generated pore network is shown in Figure 2-3 right. In this work, this pore network model was applied to the problem of diffusion and reaction occurring simultaneously in porous catalysts.

In the generation of such an irregular random pore network, the sites are distributed randomly in space at first. Then, the sites are interconnected with the neighboring sites *via* the bonds. This work showed that the average number of the bonds connected to the sites in this model, i.e., the pore interconnectivity, is an important percolation parameter and affects the pore tortuosity significantly, thus suggesting that the pore interconnectivity determined experimentally, e.g., from nitrogen sorption measurements according to Seaton (1991) [22], should be used in the generation of the pore network.

More recently, this irregular random pore network model was used by Wood and Gladden (2002) [23] to study the influence of capillary condensation on catalyst performance. This improved pore network model could simulate the change of the pore structure during the reaction, e.g., pore blocking by capillary condensation or coke formation, by removing the blocked pores in the pore network.

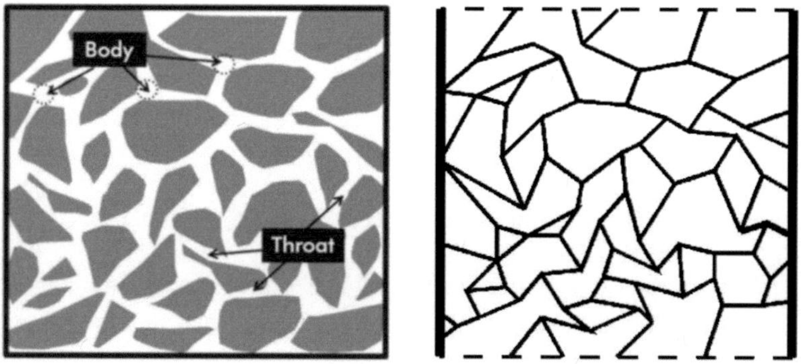

Figure 2-3: Schematic representation of porous media as pore bodies and throats (left), and its corresponding site-bond model – an irregular random pore network model (right) [23].

In these simulation models based on the irregular random pore networks ([21], [23]), diffusion was modelled according to the Fick's law, while reaction was assumed to be first order. They could not simulate the multi-component diffusion and complex reactions with general kinetics, since otherwise the computational task would be too heavy due to the irregularity and randomness of the pore network models, as well as the strong nonlinearity of the diffusion and/or reaction equations. However, in practice, the kinetics of relevant reactions often follow strong nonlinear rate equations like Langmuir-Hinshelwood or Eley-Rideal [38]. Therefore, an improved pore model, which can be applied to any common reaction kinetics, is required from the practical point of view.

2.1.4 Regular random pore network models

In contrast to irregular random pore network models, regular random pore network models as shown in Figure 2-4 have the sites representing the pore bodies distributed not randomly in the space but on the vertices of square or cube lattices. Moreover, the bonds in the lattices representing the pore throats have the same length. Same as irregular random pore network models, the bonds are distributed randomly in the regular random pore network models, and the average number of the bonds connected to the sites is equal to the pore interconnectivity determined experimentally. Therefore, the regular random pore network models can describe all the important features of the complex pore space of porous media in the same way as irregular random pore network models, but their construction is much easier, and the simulation of diffusion and reaction in such pore network models requires much less computational work and memory owing to the regularity.

In 1997, Rieckmann and Keil [24] used such a regular random pore network model – a three-dimensional cubic micro-macro pore network model shown in Figure 2-4 right to simulate the transport and reaction in the bimodal pore system of a pelletized catalyst. In this model, the bonds representing the micro- or macropores can be distributed according to a defined structure, e.g., with the macropores extending throughout the entire network (thick lines in Figure 2-4). Multicomponent transport in a single pore of the pore network was modeled by the dusty-gas model (DGM) [33], which could combine the contributions of Knudsen, molecular and viscous fluxes. Moreover, this network model can be applied to any common reaction kinetics. For example, it was applied successfully to model the deactivation of a pelletized ZSM-5 catalyst due to coke formation by removing the blocked pores in the pore network [24].

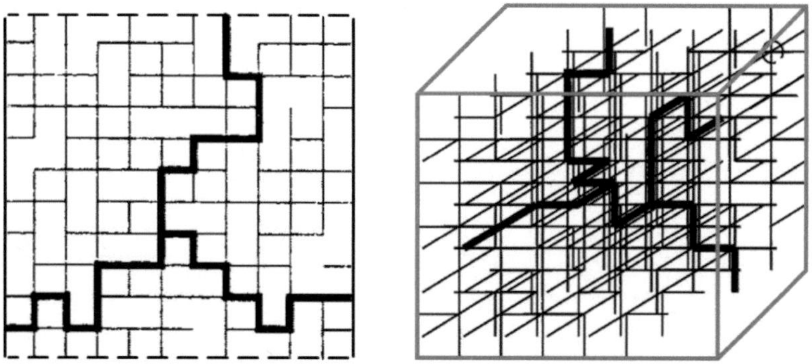

Figure 2-4: 2D (left) and 3D (right) regular random pore network models for porous media adopted from [24]. Fine and thick bonds represent the micropores (or mesopores) and macropores, respectively. The macropores extend throughout the entire network.

As an extension of the model described by Wood & Gladden (2002) [23], Wood et al. (2002) [25] used the regular random pore network model of Rieckmann & Keil [24] to simulate thiophene hydrogenation over a Co–Mo/Al_2O_3 porous catalyst. Compared to the previous model presented in [23], it can simulate multicomponent diffusion and reactions represented by any type of kinetic expression. Multicomponent diffusion involving molecular (or bulk) diffusion, Knudsen diffusion and convection of multicomponent vapors was also modeled by the dusty-gas model as in the model of Rieckmann & Keil [24].

In 2008, Chen et al. [27] used a regular three-dimensional pore network model to simulate transport and separation of binary gaseous mixtures in amorphous microporous membranes. Compared to Rieckmann & Keil's model [24], the contributions of hindered (activated Knudsen diffusion) and Knudsen diffusion as well as the viscous flux were considered in modeling of multicomponent transport of gaseous mixtures in micropores. The result of this work showed that the model was able to successfully predict the single-gas permeances and the ideal selectivity of a silicon-carbide membrane for a helium-argon system.

2.1.5 Pore-Cor network models

Laudone et al. (2008) [28] used a Pore-Cor network model to simulate liquid diffusion in a porous medium like soil, which could consider that the pore bodies (sites) had various pore volumes. The Pore-Cor network model was generated using a network simulator 'Pore-Cor' developed by the Environmental and Fluid Modeling Group at University of Plymouth, UK.

In this model, the pore structure of a porous medium is represented as a series of identical interconnected unit cells with periodic boundary conditions. Each unit cell comprises an array of 1000 nodes (10×10×10) equally spaced in a Cartesian cubic-close-packed array. Cubic pores (pore sites) with various sizes are positioned with their centers at each node, and are connected by other smaller cylindrical pores (pore bodies) in each Cartesian direction. Such a typical unit cell is shown in Figure 2-5.

The work presented in [28] showed that the Pore-Cor network model could generate pore structures which have porosity and percolation properties that closely match experimental mercury intrusion porosimetry data of real porous materials. It was applied successfully to investigate the effects of the geometry of porous materials on the diffusion process of liquid in porous materials. Moreover, this model has been previously used to model a range of materials such as soil and paper coating [29]-[30], and showed good simulation ability.

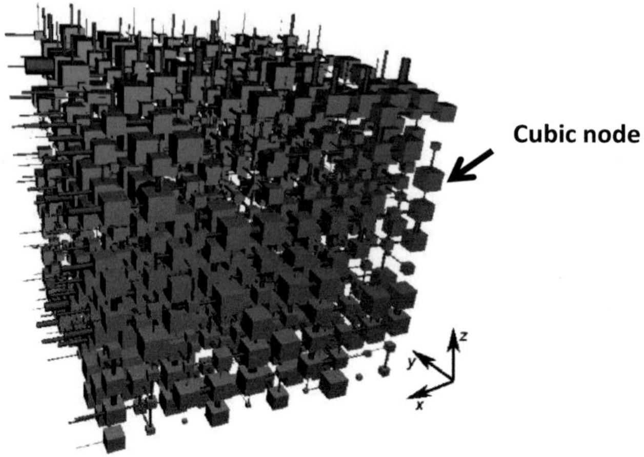

Figure 2-5: Three-dimensional representation of a Pore-Cor unit cell with 10×10×10 nodes, adopted from [28].

2.1.6 Voronoi network models

In 2011, Gostick [31] presented a random pore network model based on Delaunay and Voronoi tessellations – a Voronoi network model (Figure 2-6 right) for a fibrous gas diffusion layer (GDL) applied for fuel cell electrodes, since the Voronoi network closely resembled a fiber network and turned out to be a powerful tool for generating fibrous structures [32]. In contrast to previous pore network models, this model can describe the anisotropic structure of the fibrous gas diffusion layers.

In the construction of such a network, the first step is the random distribution of base points in space. Each of these points represents the location of a pore (black points in Figure 2-6 left). The connections between pores are defined by performing a Delaunay tessellation (the black network) on these base points. The next step is to perform a Voronoi tessellation (the red network) based on the Delaunay tessellation to define the fiber locations. This result of such 2D and 3D Voronoi networks is shown in Figure 2-6 left and right, respectively. This model not only allows a random distribution of pores in space but provides a full geometric description of pore and throat sizes, connectivity and path lengths related the fiber properties, as well as fiber backbone information. In the work of Gostick [31], this model was used to predict capillary and transport properties of gas diffusion layers and promising results were obtained.

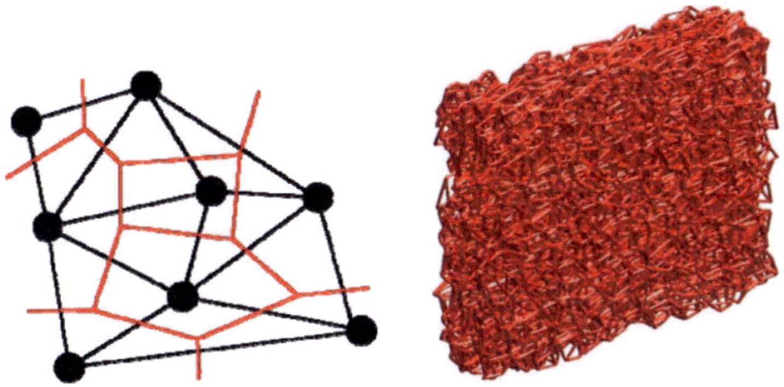

Figure 2-6: 2D (left) and 3D (right) Voronoi networks for fibrous gas diffusion layers. Red bonds represent the fibers. Diagrams are adopted from [31].

2.1.7 Summary

Table 2-1 shows the comparison of the pore network models introduced above including the ability to describe the real pore structure of porous media and computational requirement in the simulation.

Compared to other pore network models, the parallel pore networks have the lowest computational requirement. However, they fail to describe the pore structure of porous media with highly interconnected pores. The Bethe pore networks can describe the interconnectivity of the pores, but their infinite branches are not reconnected, that cannot agree with the real pore structure of most porous media, which have generally highly interconnected pores. Thus, they have limited application, only for, e.g., highly disordered porous media, but with a small pellet size.

Table 2-1: Comparison of pore network models for porous media available in literature

Pore networks	**Best for which media**	**Simulation Ability**	**Computational Requirement**
Parallel	Not highly interconnected pores	--	--
Bethe	Small amorphous pellets	-	-
Irregular random	Amorphous, disordered	++	+
Regular random	Amorphous, disordered	+	-
Pore-Cor	Amorphous (e.g., soil)	+	-
Voronoi	Fibrous (e.g., GDLs)	++	++

--: very low; -: low; +: high; ++: very high.

In contrast to the parallel and Bethe pore networks, the irregular and regular pore network models can well simulate the real pore structures of most amorphous porous media with highly interconnected pores. Owing to the regularity, the regular pore network models have much lower computational requirement than the irregular pore network models. Therefore, they are more applicable to commercial porous catalysts with complex transport mechanisms in pores and strongly nonlinear reactions.

Different from the pore models mentioned above, the Pore-Cor and Voronoi pore network models enjoy applicability to porous media with special pore structures. The Pore-Cor pore network models are developed for the porous media like soil, in which the diffusion and concentration change of fluids are affected significantly by the size of the pore bodies, i.e., the sites in the network models. Since they have a similar structure as the regular pore network models, they also have low computational requirement. The Voronoi pore network models have a more complex structure to simulate the pore structure of fibrous media like GDLs. Therefore, they can simulate the anisotropic transport properties of the GDLs, but require higher computational memory and ability.

Unfortunately, all the models presented above cannot describe the polycrystalline and anisotropic pore structures of the ZSM-5 zeolite catalysts as shown in Figure 1-1 and Figure 1-3. However, as mentioned above, their polycrystalline and anisotropic structures have shown significant effect on their separation and catalytic performance. Thus, a realistic pore model, which can simulate polycrystalline and anisotropic structure, is needed for a rational simulation-assisted design of such media.

2.2 Gas transport in porous media

Gas transport in a porous medium can be influenced by factors like the pore structure of the porous medium (e.g., pore size), the molecule size of the diffusing gas species, the temperature and the pressure difference. As shown in Figure 2-7, gas transport in a pore occurs to the following mechanisms with decreasing pore size:

a) Molecular diffusion;
b) Convective flow, i.e., viscous flow;
c) Knudsen diffusion;
d) Surface diffusion;
e) Single-file diffusion;
f) Activated Knudsen diffusion (or hindered diffusion).

The last three transport mechanisms (surface diffusion, single-file diffusion and activated Knudsen diffusion) belong to the class of configurational diffusion, which is characterized by its activated nature different from other transport mechanisms. They generally occur in micropores due to their narrow pore space.

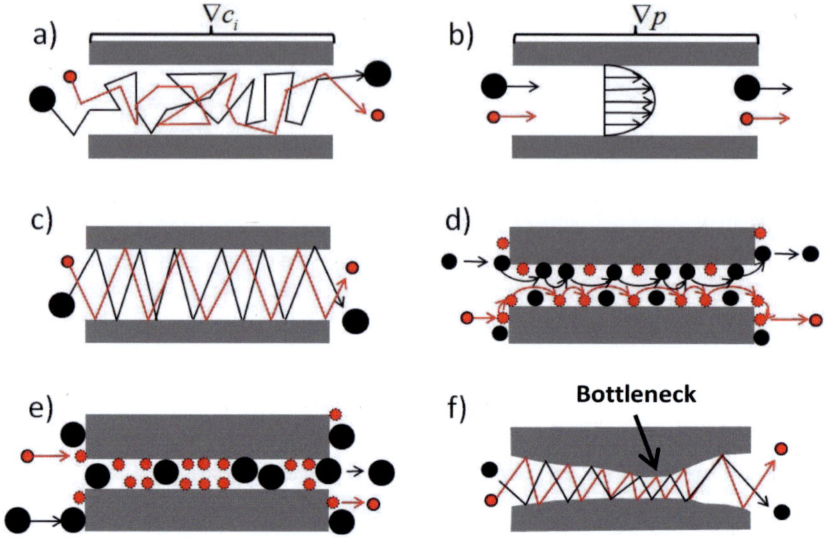

Figure 2-7: Various transport mechanisms in porous media: a) Molecular diffusion driven by a concentration gradient; b) Viscous flow driven by a pressure gradient; c) Knudsen diffusion; and configurational diffusions in micropores: d) Surface diffusion; e) Single-file diffusion; f) Activated Knudsen diffusion.

2.2.1 Molecular diffusion

As shown in Figure 2-7 a), molecular diffusion (or bulk diffusion) is driven by collisions between gas molecules in motion under a concentration gradient of gas species. Under low to moderate pressures (e.g., below 40 bar), the equation of Fuller, Shettler and Giddings [36] can be used to calculate the binary molecular diffusion coefficient between species *i* and *j*:

$$D_{i,j}^{M} = \frac{1.013 \cdot 10^{-7} T^{1.75} \left(\frac{M_i + M_j}{M_i \cdot M_j} \right)^{1/2}}{p \left[\left(\sum v_{\mathrm{diff},i} \right)^{1/3} + \left(\sum v_{\mathrm{diff},j} \right)^{1/3} \right]^2}, \qquad \text{Eq. 2—1}$$

T : temperature, K; p : total pressure, bar; M : molar mass of gas species, kg mol^{-1}; v_{diff} : diffusion volume.

Derived from the Maxwell-Stefan equations:

$$-\nabla c_i = \sum_{j \neq i}^{n} \frac{c_j J_i^M - c_i J_j^M}{c_{tot} \cdot D_{i,j}^M}, \qquad \text{Eq. 2—2}$$

∇c_i : the concentration gradient of species *i*, mol m^{-4}; c_{tot} : the total concentration, mol m^{-3}, the multicomponent fluxes of n-1 gas species due to molecular diffusion can be calculated by the following matrix:

$$\mathbf{J}^{\mathbf{M}}_{(\mathbf{n}-1)} = -\left(\mathbf{B}^{\mathbf{M}}_{(\mathbf{n}-1)\times(\mathbf{n}-1)}\right)^{-1} \nabla \mathbf{c}, \text{ with} \qquad \text{Eq. 2—3}$$

$$\begin{cases} B_{i,i}^M = \dfrac{c_i}{c_{tot} \cdot D_{i,n}^M} + \displaystyle\sum_{\substack{j \neq i \\ j=1}}^{n} \dfrac{c_j}{c_{tot} \cdot D_{i,j}^M} \\ B_{i,j}^M = \dfrac{c_i}{c_{tot} \cdot D_{i,n}^M} - \dfrac{c_i}{c_{tot} \cdot D_{i,j}^M} \end{cases}, \qquad \text{Eq. 2—4}$$

while the flux of the nth gas species is calculated from the obtained fluxes of other n-1 gas species using Eq. 2—2.

2.2.2 Viscous flow

Different from molecular diffusion, the driving force for viscous flow is the absolute pressure gradient (see Figure 2-7 b)). Under the assumptions that:

- the flow in the pore is laminar,
- the gas is an incompressible Newtonian fluid, and
- the cylindrical pore has a constant cross section,

the viscous flux in the pore can be calculated by the Poiseuille equation:

$$J_i^V = \frac{r_p^2 \rho_g}{8\eta} \nabla p_i, \qquad \text{Eq. 2—5}$$

r_p: represents the pore radius, m; ρ_g is the gas density, kg m^3; η is the dynamic viscosity of the gas, Pa s.

2.2.3 Knudsen diffusion

As shown in Figure 2-7 c), since the pore is so narrow that its pore diameter is comparable to or smaller than the mean free path of the gas molecules, the gas molecules diffuse through the pore mainly *via* the collisions with the pore wall. Therefore, different from the molecular diffusion, the molecules do not interact with one another in the Knudsen diffusion regime.

The diffusivity for Knudsen diffusion can be calculated with the following equation, which is derived from the kinetic theory of gases [37]:

$$D_i^K = \frac{d_p}{3} \sqrt{\frac{8RT}{\pi M_i}}, \qquad \text{Eq. 2—6}$$

where d_p represents the pore diameter, m.

Thus, the Knudsen flux in the pore can be calculated by the following equation:

$$J_i^K = \underbrace{\frac{d_p}{3} \sqrt{\frac{8RT}{\pi M_i}}}_{D_i^K} \cdot \nabla c_i, \qquad \text{Eq. 2—7}$$

where ∇c_i represents the concentration gradient of species i.

2.2.4 Surface diffusion

In a narrow pore with a pore diameter comparable to the kinetic diameter of the gas molecules, the gas molecules diffuse in adsorbed state on the pore wall as shown in Figure 2-7 d). In order to model such a diffusion mechanism (i.e., surface diffusion), Krishna (1990) [18] proposed the generalized Maxwell-Stefan surface diffusion model (SDM). SDM considers the chemical potential gradient $\nabla\mu$ as the driving force. Its mathematical expression for a multicomponent system is represented as follows:

$$-\frac{\theta_i}{RT}\nabla\mu_i=\sum_{j=1}^{n}\frac{q_jJ_i^S-q_iJ_j^S}{q_{sat,i}\cdot q_{sat,j}\cdot\rho\cdot D_{i,j}^S}+\frac{J_i^S}{q_{sat,i}\cdot\rho\cdot D_i^S}\cdot \qquad \text{Eq. 2—8}$$

where θ_i is the coverage of species i on the medium; q_i represents the loading of species i on the medium, mol kg^{-1}; $D_{i,j}^S$ and D_i^S are mutual Maxwell-Stefan diffusion coefficient and corrected diffusion coefficient, respectively; ρ is the density of the medium. The mutual Maxwell-Stefan diffusion coefficient describes the friction between the adsorbed molecules, while the corrected diffusion coefficient represents the friction between the adsorbed molecules and the solid surface of the pore wall. The corrected diffusion coefficient can be determined experimentally by singe-gas permeation studies [39], e.g., in a Wicke-Kallenbach cell (for more details, see Section 4.2). The mutual Maxwell-Stefan diffusion coefficient cannot easily be determined experimentally but can be estimated by the following equations [18], [40]:

$$q_iD_{j,i}^S=q_jD_{i,j}^S=\left(q_jD_{i,i}^S\right)^{q_i/(q_i+q_j)}\cdot\left(q_iD_{j,j}^S\right)^{q_j/(q_i+q_j)}, \qquad \text{Eq. 2—9}$$

$$D_{i,i}^{S} = \frac{\theta_i}{\dfrac{1}{D_{self,i}^{S}} - \dfrac{1}{D_i^{S}}},$$ Eq. 2—10

where $D_{i,i}^{S}$ and $D_{self,i}^{S}$ represent the Maxwell-Stefan self-exchange diffusivity and self-diffusivity, respectively. Eq. 2—10 shows that the self-diffusivity must be known for the calculation of the mutual Maxwell-Stefan diffusivity. The self-diffusivity can be measured by nuclear magnetic resonance pulsed field gradient spectroscopy (PFG-NMR) [41]-[42] or quasielastic neutron scattering (QENS) [42], or estimated by molecular dynamics simulation (MD) [43]. When the self-diffusivities are not available, Krishna [18] calculated the mutual Maxwell-Stefan diffusivities by replacing the Maxwell-Stefan self-exchange diffusivities in Eq. 2—9 with the corrected diffusivities:

$$q_i D_{j,i}^{S} = q_j D_{i,j}^{S} = \left(q_j D_i^{S}\right)^{q_i/(q_i+q_j)} \cdot \left(q_i D_j^{S}\right)^{q_j/(q_i+q_j)}.$$ Eq. 2—11

The chemical potential gradient in Eq. 2—8 is related to the surface coverage gradient by a thermodynamic factor Γ:

$$\frac{\theta_i}{RT}\nabla\mu_i = \sum_{j=1}^{n}\Gamma_{i,j}\nabla\theta_i.$$ Eq. 2—12

2.2.4.1 Sorbate molecules with similar saturation loadings

If the sorbate molecules have similar saturation loadings, which allows the average saturation loading $\overline{q}_{sat}$ to replace all the saturation loadings of species in Eq. 2—8 for simplicity, the following expression in form of matrix and vector was obtained by Krishna and co-workers (1990) [18] to calculate the surface fluxes *via* substituting Eq. 2—12 into Eq. 2—8:

$$\mathbf{J}^S = -\rho \cdot q_{sat} \mathbf{D}^S \nabla \boldsymbol{\theta}, \qquad \text{Eq. 2—13}$$

where

$$\mathbf{D}^S = \mathbf{B}^{-1}\boldsymbol{\Gamma}, \text{ with} \qquad \text{Eq. 2—14}$$

$$B_{i,j} = \begin{cases} -\dfrac{\theta_i}{D^S_{i,j}}, & \text{for} \quad i \neq j; \\ \dfrac{1}{D^S_i} + \sum\limits_{\substack{j \neq i \\ j=1}}^{n} \dfrac{\theta_j}{D^S_{i,j}}, & \text{for} \quad i = j. \end{cases} \qquad \text{Eq. 2—15}$$

If the multicomponent adsorption isotherm can be modeled with the extended Langmuir equations, the thermodynamic factor between species i and j is rewritten with the follow equations:

$$\Gamma_{i,j} \equiv \theta_i \frac{\partial \ln p_i}{\partial \theta_j} = \frac{\theta_i}{p_j} \frac{\partial p_i}{\partial \theta_j} = \begin{cases} \dfrac{\theta_i}{1 - \sum\limits_{j=1}^{n} \theta_j} & i \neq j \\ 1 + \dfrac{\theta_i}{1 - \sum\limits_{j=1}^{n} \theta_j} & i = j \end{cases} \qquad \text{Eq. 2—16}$$

where n is the species number in the gas mixture, $i, j = 1, 2, \cdots n.$.

2.2.4.2 Sorbate molecules with similar saturation loadings

In 2000, Kapteijn et al. [44] found that since the molecules of hydrocarbons (e.g., ethane/methane and propane/methane) have quite different saturation loadings, e.g., on silicalite-1 membranes, the simulation of gas separation of ethane/methane and propane/methane mixtures could not agree with the experimental data well, if equal saturation loadings were assumed for all species in the mixture. Therefore, they suggested different saturation loadings of species in SDM, and to used the ideal adsorbed solution theory

(IAST) proposed by Myers and co-workers [45] for the calculation of the thermodynamic factors. A new expression in form of matrix and vector to calculate the surface fluxes can be obtained:

$$\mathbf{J}^S = -\rho \cdot \mathbf{D}'^S \nabla \boldsymbol{\theta}, \qquad \text{Eq. 2—17}$$

where

$$\mathbf{D}'^S = \mathbf{B}'^{-1} \boldsymbol{\Gamma}, \qquad \text{Eq. 2—18}$$

with $$B'_{i,j} = \begin{cases} -\dfrac{q_i}{q_{sat,i} q_{sat,j} D^S_{i,j}}, & i \neq j; \\ \dfrac{1}{q_{sat,i} D^S_i} + \sum\limits_{\substack{j \neq i \\ j=1}}^{n} \dfrac{q_j}{q_{sat,i} q_{sat,j} D^S_{i,j}}, & i = j. \end{cases} \qquad \text{Eq. 2—19}$$

Using the IAST for multicomponent adsorption, the thermodynamic factors between species in the matrix $\boldsymbol{\Gamma}$ have no analytic solutions as shown in Eq. 2—16 from the extended Langmuir equations, since the loadings of species have no analytic relations with the partial pressures of species. From the IAST, the molar fraction of species i in all adsorbed molecules on the solid surface of the medium (x_i) can be obtained from the partial pressures of species in the gas phase (p_i) by solving the following system of n nonlinear equations [45]:

$$\begin{bmatrix} \sum\limits_{i=1}^{N} x_i - 1 \\ \vdots \\ q_{sat,i} \ln\left(1 + K_i \dfrac{p_i}{x_i}\right) - q_{sat,i+1} \ln\left(1 + K_{i+1} \dfrac{p_{i+1}}{x_{i+1}}\right) \\ \vdots \\ q_{sat,n-1} \ln\left(1 + K_{n-1} \dfrac{p_{n-1}}{x_{n-1}}\right) - q_{sat,n} \ln\left(1 + K_{n-1} \dfrac{p_n}{x_n}\right) \end{bmatrix} = \mathbf{0}, \qquad \text{Eq. 2—20}$$

where n is the number of species in the mixture, K_i represents the adsorption equilibrium constant of species *i*, which is determined from the single-component adsorption isotherm *via* fitting with the Langmuir equation.

Using the IAST, the coverage of species i (θ_i) can be calculated from the molar fraction of species i in all adsorbed molecules (x_i) with the following equation:

$$\theta_i = \frac{q_i}{q_{sat,i}} = \frac{q_{tot} x_i}{q_{sat,i}} = \frac{x_i}{q_{sat,i} \sum_{m=1}^{N} \frac{x_m}{q_m^0}}, \qquad \text{Eq. 2—21}$$

where q_{tot} and q_m^0 represent the total loading of adsorbed molecules and the loading of species m in the absence of other species under the gas pressure of p_i, respectively.

Finally, the partial derivative ($\frac{\partial p_i}{\partial \theta_j}$) in the Eq. 2—16 to calculate the thermodynamic factor can be derived from the partial pressures p_i and the coverages of species (θ_i) obtained in Eq. 2—21.

2.2.5 Single-file diffusion

As shown in Figure 2-7 e), since the larger adsorbed molecules (black) have a kinetic diameter close to the pore diameter, the smaller adsorbed molecules (red) cannot overtake them. Thus, mutual passage of molecules is excluded (i.e. $D_{i,j}^s \rightarrow 0$), i.e., the transport rate of molecules in the pore is controlled by the transport rate of the larger molecules. This effect is significant particularly when the loading of the adsorbed molecules on the pore wall is high. This special transport mechanism is named single-file diffusion

(SFD). A detailed introduction about SFD can be found in Chapter 5 of the book – *Diffusion in Nanoporous Materials* written by Kärger et al. [46].

Since *p*-, *o*- and *m*-xylenes have a diameter of 5.8, 6.8, and 6.8 Å, respectively, the effect of SFD was observed in the separation of xylene isomers with silicalite-1 membranes in experiments by Baertsch et al. [47] and Xomeritakis et al. [48]. They have used SFD to explain the absence of any separation on gas mixtures of xylene isomers with silicalite-1 membranes, when the loading of xylene isomers in silicalite-1 membranes is high. Moreover, the simulation results of van den Broeke et al. [49] showed that the single-file diffusion model (SFDM) could predict the separation factor for the binary mixture of *n*-butane and *i*-butane through a Silicalite-1 membrane well. Compared to surface diffusion, for single-file diffusion, the elements of the matrix $\mathbf{B}^{-1}$ in Eq. 2—14 are:

$$B_{i,j}^{-1} = \begin{cases} 0 & \text{for} \quad i \neq j; \\ D_i^S & \text{for} \quad i = j. \end{cases} \qquad \text{Eq. 2—22}$$

2.2.6 Activated Knudsen diffusion

As shown in Figure 2-7 f), the micropores of amorphous microporous media like silica generally do not have a constant cross section, but have the narrow 'bottleneck' somewhere. Therefore, the gas molecules have to overcome an energy barrier due to the 'bottleneck' when diffusing through the pore.

Xiao and Wei [34] described such activation diffusion in a micropore as the product of a shape factor *g*, the velocity of the molecule and an exponential term, based on the conventional Knudsen diffusion described in Section 2.2.3. Therefore, such diffusion is named activated Knudsen diffusion with the diffusion coefficient that:

$$D_i^{AK} = g \cdot \underbrace{\frac{d_p}{3}\sqrt{\frac{8RT}{\pi M_i}}}_{D_i^K} \cdot \exp\left\{\frac{-E_a^g}{RT}\right\}, \qquad \text{Eq. 2—23}$$

where g is a geometrical factor that accounts for the porosity and the tortuosity of the pore, and E_a^g is the activation energy to overcome the energy barrier of the diffusion.

2.2.7 Summary

As mentioned above, the pore size plays an important role in the gas transport in the porous media. According to the IUPAC classification [50], the pores in porous media are generally classified into macro-, meso-, and micropores as shown in Table 2-2. The macro-, meso-, and micropores typically have pore size larger than 50 nm, between 2 and 50 nm, and smaller than 2 nm, respectively.

As concluded in Table 2-2 and Figure 2-8, the gas transport mechanisms in macropores can be molecular diffusion under a concentration gradient and viscous flow under a pressure gradient. Sometimes, Knudsen diffusion plays a role. In general, the dominant transport mechanism in a macropore is molecular diffusion under a significant concentration gradient, while viscous flow is dominant when the pressure gradient in the pore is significant, e.g., a rapid gas volume change occurs due to the reaction in the pore, or there is a significant pressure difference over a porous membrane.

In a mesopore, Knudsen diffusion generally has the main contribution on the gas transport. Sometimes, particularly when the pore has a size close to the micropores, surface diffusion, single-file diffusion and activated Knudsen diffusion will also play a role. Due to the narrow space in the micropores, in general, only surface diffusion, single-file diffusion and activated Knudsen diffusion can contribute the transport. Sometimes, Knudsen diffusion plays a role. Finally, we can find that which transport mechanism is dominant depends on the pore structure and adsorption property of the porous media, as well as on the properties of the gas molecules.

Table 2-2: Pore types according to IUPAC classification and corresponding gas transport mechanisms in pores.

Pore types	**Macropore**	**Mesopore**	**Micropore**
Pore size	d > 50 nm	2 nm < d < 50 nm	d < 2 nm
Gas transport mechanisms	• Molecular diffusion • Viscous flow • Knudsen diffusion	• Knudsen diffusion • Surface diffusion • Single-file diffusion • Activated Kn. diffusion	• Knudsen diffusion • Surface diffusion • Single-file diffusion • Activated Kn. Diffusion

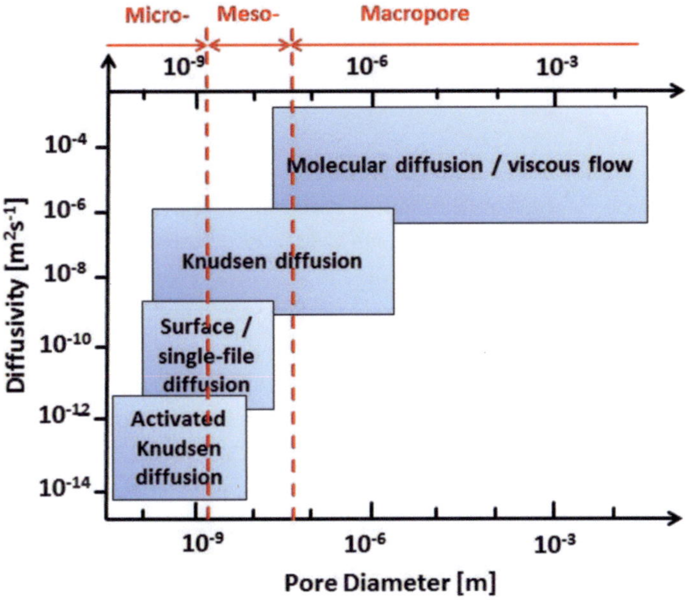

Figure 2-8: Gas transport mechanisms in pores as a function of pore diameter.

In order to simulate multicomponent gas transport in the pores, where viscous flow, molecular diffusion, Knudsen diffusion, and surface diffusion occur simultaneously, the dusty-gas model (DGM) was used by Rieckmann and Keil (1997) [24] and Wood et al. (2002b) [25] to combine the contributions of these transport mechanisms. In DGM, the pore walls are considered as consisting of giant molecules ('dust') distributed in space, while the following expression in form of matrix and vector was obtained by Rieckmann and Keil (1997) [24] to calculate the combined molecular and Knudsen flux:

$$\mathbf{J}^D = -\mathbf{D}^{-1}\nabla\mathbf{c}\,, \qquad \text{Eq. 2—24}$$

$$\text{with} \quad D_{i,j} = \begin{cases} -\dfrac{x_i}{D_{i,j}^M}, & \text{for} \quad i \neq j; \\ \dfrac{1}{D_i^K} + \sum\limits_{\substack{h \neq i \\ h=1}}^{n} \dfrac{x_h}{D_{i,h}^M}, & \text{for} \quad i = j. \end{cases}$$ Eq. 2—25

The combination of the contribution of different transport mechanisms in the model of Rieckmann and Keil (1997) [24] is visualized by an electric analogue in Figure 2-9. The total flux is the sum of the combined molecular and Knudsen flux $\mathbf{J}^D$, the viscous flux $\mathbf{J}^V$, and the surface flux $\mathbf{J}^S$:

$$\mathbf{J}_{\mathbf{tot}} = \mathbf{J}^D + \mathbf{J}^V + \mathbf{J}^S .$$ Eq. 2—26

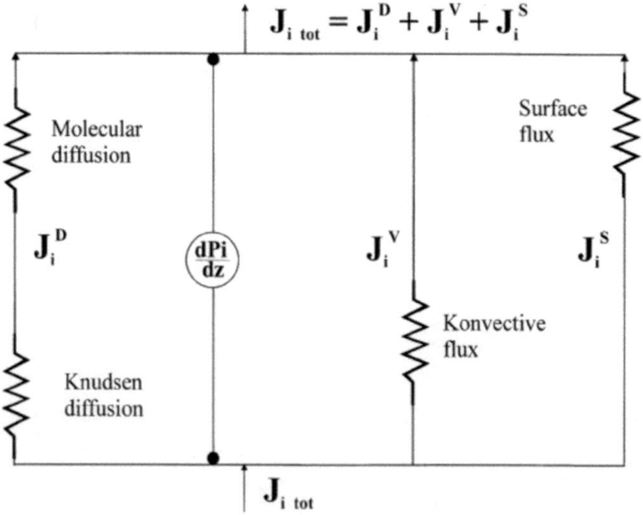

Figure 2-9: An electric analogue to describe the combination of transport mechanisms, adopted from Rieckmann and Keil (1997) [24].

3 Crystallite-pore network model for polycrystalline microporous media

Based on the previous pore network models introduced in the last chapter, a crystallite-pore network model (CPNM) was developed in this dissertation to simulate the pore structure of the anisotropic polycrystalline microporous media such as zeolite ZSM-5 membranes or layers. Combining the transport and kinetic models with this pore network model, a simulation model is obtained to simulate transport and reaction of multicomponent gas mixtures in anisotropic polycrystalline microporous media. The applications of the CPNM on core-shell catalysts and catalytic zeolite membranes will be shown in Chapters 6 and 7, respectively.

3.1 Development of crystallite-pore network model

A polycrystalline microporous medium such as a ZSM-5 zeolite membrane is represented schematically in Figure 3-1. The microporous crystallites, in which intracrystalline micropores exist, are surrounded by the intercrystalline mesopores (i.e. defects such as gaps, pinholes and cracks). In order to obtain defect-free membranes with excellent separation performance, the defects should be controlled at a low concentration in membrane preparation e.g. *via* treatments such as filling the defects by a post-synthetic coking treatment [52], etc.

Different from amorphous microporous materials such as silica membranes, many polycrystalline microporous membranes have an anisotropic pore system with micropores of different structure and/or size in different orientation, e.g., ZSM-5 crystallites have zig-zag and straight intracrystalline micropores in *a*- and *b*-direction, respectively, as shown in Figure 1-3. As a result, there has been intensive interest by the research community to synthesize zeolite membranes with well-oriented structure in order to obtain better permeation and separation performance unavailable from the randomly oriented zeolite membranes. An example is the preparation of a *b*-oriented ZSM-5 zeolite membrane by Lai et al. [16] for gas separation, e.g., xylene isomers.

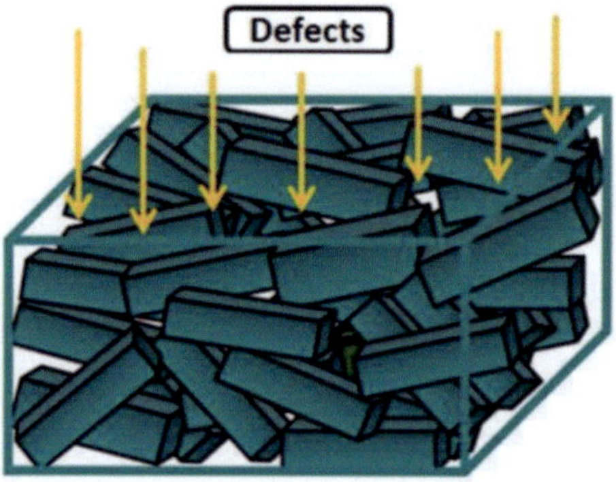

Figure 3-1: Schematic showing a polycrystalline microporous membrane with intercrystalline pores, i.e., defects.

In order to simulate the complex pore space in polycrystalline membranes as shown in Figure 3-1, a crystallite-pore network model was developed by including distributed nodes representing crystallites in the conventional cubic pore network model (see Figure 2-4). As an example, a crystallite-pore network with 8 crystallites (2×2×2) is illustrated in Figure 3-2. Considering that crystallites grow together to form the polycrystalline membrane, the adjacent nodes representing different crystallites are connected with each other *via* the interface nodes (red nodes), as shown in the exemplary crystallite-pore network. These nodes are also connected with the nodes of the intercrystalline pores (light grey nodes). In this way, the interface nodes establish a connection between the intra- and intercrystalline pore networks. Moreover, in order to simulate anisotropy of the crystallites in the polycrystalline layer, every crystallite in the pore network model is allowed to have two angles Ψ and X to describe its individual orientation, as shown in Figure 3-2 right.

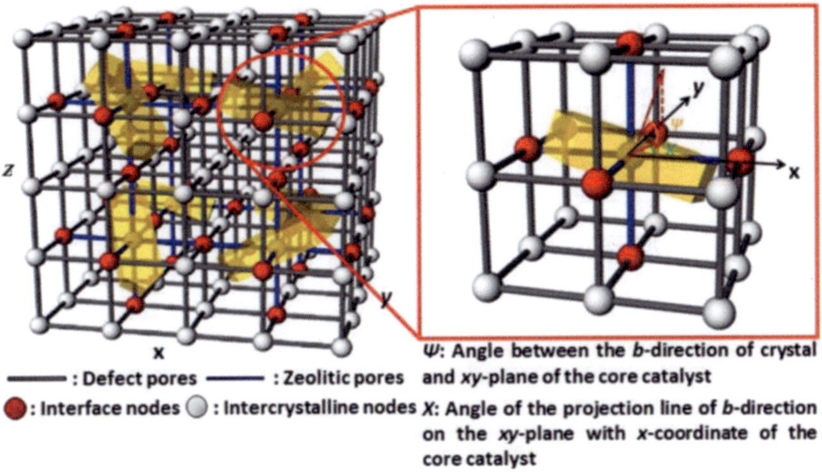

Figure 3-2: Crystallite-pore network model [53] with the size 2×2×2 crystallites³ (left). One crystallite with angles Ψ and X surrounded by the intercrystalline mesopores (right).

The connectivity of the intercrystalline mesopores in the CPNM is defined as:

$$C = \frac{\text{total number of intercrystalline pores}}{\text{total number of nodes of interfaces and intercrystalline pores}}, \qquad \text{Eq. 3—1}$$

which gives the average number of intercrystalline mesopores connected on each node of the network. The connectivity of the pores can be determined experimentally, e.g., by the method of Seaton [22]. In the CPNM, the connectivity in accordance with the experimental value is achieved by specifying a connection probability P of the intercrystalline mesopores during generation of the network. P ranges from 0 to 100 %, where 0 means all defects are isolated and 100 % all defects are connected. For all values in between, the connections between the defect nodes are of statistical nature. The connectivity generally depends on the network size as well. For small networks, the connectivity of the resulting intercrystalline pore network for a given value of P is subject to statistical variation. This can be seen in Figure 3-3 showing a plot of the connectivity C from Eq. 3—1 as a function of the connection probability P for a network with 5×5×7 nodes (i.e., 2×2×3 crystallites).

In reality, the crystallites are heavily inter-grown to reduce defects, i.e. intercrystalline pores. The crystallite-pore network model is also applicable to such heavily inter-grown polycrystalline membranes. In this case, the crystallite node on a lattice will represent an inter-grown polycrystalline layer, outside surrounded by the defect pores, but inside without defects or with isolated defects.

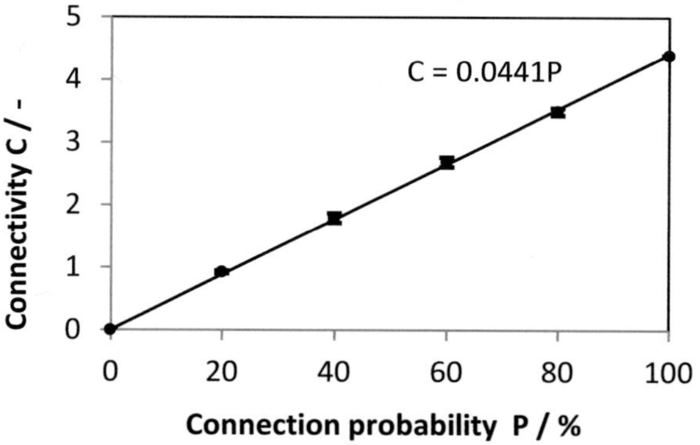

Figure 3-3. Connectivity of intercrystalline mesopores as a function of the connection probability for a network with 5×5×7 nodes (i.e., 2×2×3 crystallites). Error bars give the standard errors for five consecutive calculations, adopted from [53].

Consequently, in contrast to previous pore networks, the crystallite-pore network model is advantageous with regard to the following aspects:

- The effect of the defects can be investigated *via* varying their concentration, connectivity, and pore size;
- The effect of anisotropy of crystallites, e.g., the dependence of diffusivity on the orientation of crystallites, can be studied by varying the orientation of crystallites;
- The effect of crystallite size can be simulated;
- The effect of interaction between intra- and intercrystalline pore systems can be simulated;
- Blocking on the crystallite surface or in intracrystalline pores due to coking (catalyst deactivation) can also be simulated;
- As an extension on the network model, secondary pore networks are allowed to replace the nodes representing crystallites so that the concentration distribution of reactants and products within crystallites can be simulated.

3.2 Combination of transport and kinetic models in the CPNM

Transport and kinetic models were combined in the CPNM to simulate transport and reaction of multicomponent gas mixtures in the polycrystalline microporous media, e.g., ZSM-5 membranes or layers. The defects typically found in ZSM-5 membranes vary with the size of the crystallites and generally have a width or diameter of 2-20 nm. The Knudsen diffusion model was used to simulate diffusion in these defects, since compared to the Knudsen flow, the surface flow is negligibly small due to low loadings of adsorbed species on the pore surface of the defects, particularly at higher temperatures. The flow of species i through a single intercrystalline mesopore was calculated by:

$$N_{inter,i} = -A_{inter} J_i^K , \qquad \text{Eq. 3—2}$$

with the cross-sectional area of an intercrystalline pore: $A_{inter} = \pi \frac{d_{p,inter}^2}{4}$, and the Knudsen flux J_i^K was calculated with Eq. 2—7 in Section 2.2.3.

Besides the intercrystalline defects, most zeolites have intracrystalline pores with a diameter smaller than 1 nm, e.g., the intracrystalline pores in the *b*-direction of ZSM-5 crystallites have a diameter of 5.6 Å. Due to their narrow pore space, configurational diffusions, i.e., surface diffusion, single-file diffusion and activated Knudsen diffusion, is the dominant mechanism of transport in these micropores. It was reported by Van de Graaf et al. [39] that the generalized Maxwell-Stefan surface diffusion model (SDM) without considering activated Knudsen diffusion could successfully predict the binary permeation of ethane/methane and propane/methane through a silicalite-1 membrane (silicalite-1 has the same microstructure as ZSM-5). Therefore, without considering the activated Knudsen diffusion, SDM was used in the simulation models based on the CPNM to simulate diffusion of related gases in direct DME synthesis (i.e., CO, H_2, CO_2, H_2O, methanol, DME with a molecular size range of 2-4 Å) in the intracrystalline micropores of ZSM-5, while the single-file diffusion model (SFDM) described the diffusion of xylene isomers with molecular sizes in the range of 5.8-6.5 Å. Moreover, only the flux in *b*-direction was considered, since diffusion measurements with pulsed field gradient-nuclear magnetic resonance (PFG-NMR) spectroscopy by Hong et al. [41] indicated that the diffusivities of molecules in *a*- and *c*-direction are much lower than those in *b*-direction. Therefore, the flow of species i through one intracrystalline pore can be calculated from the surface flux of species i (J_i^S) with the following equation:

$$N_{intra,i} = -\alpha A_{intra} J_i^S , \qquad \text{Eq. 3—3}$$

where the cross-sectional area of the intracrystalline pore $A_{intra} = \pi \frac{d^2_{p,intra}}{4}$ and the factor α describes the influence of the crystallite orientation on the transport in the crystallites. The angles ψ and χ shown in Figure 3-2 right represent the angle of the *b*-direction of a crystallite with the *xy*-plane and the angle of the projection line of *b*-direction on the *xy*-plane with the *x*-coordinate, respectively. The angle distributions of ψ and χ can be determined experimentally, e.g., *via* analyzing scanning electron microscope (SEM) images or X-ray diffraction (XRD) pole-figures. This latter method has been applied by Lovallo et al. [54] to determine the crystal orientation of MFI-type membranes. Based on these two parameters, the factor α representing the distribution of the flow in different coordinates can be calculated as follows:

$$\alpha = \begin{cases} \cos^2\psi \cos^2\chi & x-\text{coordinate}, \\ \cos^2\psi \sin^2\chi & y-\text{coordinate}, \\ \sin^2\psi & z-\text{coordinate}. \end{cases} \qquad \text{Eq. 3—4}$$

Rieckmann and Keil [24] pointed out that the porosity in a pore network without micro and macro scaling is orders of magnitude less than that of a real porous catalyst. Here, the porosity represents the ratio of the pore volume and the network volume. If the pore size and length do not vary, the porosity is identical to the pore concentration. In order to obtain realistic prediction results, the flows in the inter- and intracrystalline pores therefore were scaled by the factors ς_{inter} and ς_{intra} following Rieckmann and Keil's method [24]:

$$N'_{inter,i} = N_{inter,i} \frac{V_{inter,\ memb}}{V_{inter,net}} = N_{inter,i} \cdot \varsigma_{inter}, \qquad \text{Eq. 3—5}$$

$$N'_{intra,i} = N_{intra,i} \frac{V_{intra,\ memb}}{V_{intra,net}} = N_{intra,i} \cdot \varsigma_{intra}. \qquad \text{Eq. 3—6}$$

Due to high specific internal surface area of microporous crystallites, it can be considered that the most catalytic active sites are distributed within the crystallites. Thus, it can be assumed in the crystallite-pore network model that the reaction takes place only in the intracrystalline micropores within the ZSM-5 crystallites. According to the component material balance of species i for the node representing a crystallite, it holds:

$$\sum_{\substack{ip \\ ip \in I}} N'^{ip}_{intra,i} + m_{cry} (\mathbf{v} \cdot \mathbf{r})_i = \mathbf{0} \qquad \text{Eq. 3—7}$$

where I represents all the pores connected at the node representing a crystallite and $(\mathbf{v} \cdot \mathbf{r})_i$ is the reaction rate of species i.

Under the assumption of zero coverage gradients within the microsized crystallites, the vector of the reaction rates $\mathbf{r}$ in a crystallite can be calculated with the partial pressures or coverages of species at the node representing this crystallite:

$$\mathbf{r} = \mathbf{r}(p_{cry,i} / \theta_{cry,i}, i = 1, \cdots, n). \qquad \text{Eq. 3—8}$$

It is assumed that no reaction takes place in the intercrystalline mesopores. According to the component material balances for nodes representing intercrystalline mesopores and interfaces, the flows of species entering the node must therefore be equal to the flows of species leaving the node. Thus, the following equation holds at every node of intercrystalline pores and interfaces:

$$\sum_{\substack{ip \\ ip \in I}} N'^{ip}_i = \mathbf{0} \cdot \qquad \text{Eq. 3—9}$$

A no-flow boundary condition (a symmetry condition) was assumed for the four surfaces of the network (i.e., those perpendicular to the membrane surface), namely,

$$N_i' = \mathbf{0}\,.$$ Eq. 3—10

Finally, assembling all the above equations for nodes in the network according to the component material balances, a large system of nonlinear equations is obtained:

$$\mathbf{F}(\mathbf{c}) = \mathbf{0}\,.$$ Eq. 3—11

3.3 Implementation of simulation models in Matlab®

The simulation models based on the CPNM reported in the last subsections for the DME synthesis in the core-shell catalyst and xylene isomerization in a catalytic zeolite membrane were implemented in Matlab®. The simulation codes in Matlab® consist of a main routine file and a function file. In the function file named by 'myfun_CPNM', the large system of nonlinear equations in Eq. 3—11 is coded. Figure 3-4 shows its overall algorithm, in which the part of the function file is identified using the dotted line. The main routine file is divided into four parts as follows, and each part represents a step in the calculation:

1) Construction of a crystallite-pore network, if necessary, the constructed pore network can be displayed;
2) Calculation of the concentration distribution in the CPNM *via* solving the large nonlinear equation system in the function file;
3) After checking the obtained concentration distribution *via* the component balance at every node, as well as the C- and H-balance, calculating the results (e.g., conversion of reactants and selectivities to products) from the obtained concentration distribution in Step 2, as well as the concentration profiles in the CPNM;
4) Display of the species concentration in the pore network.

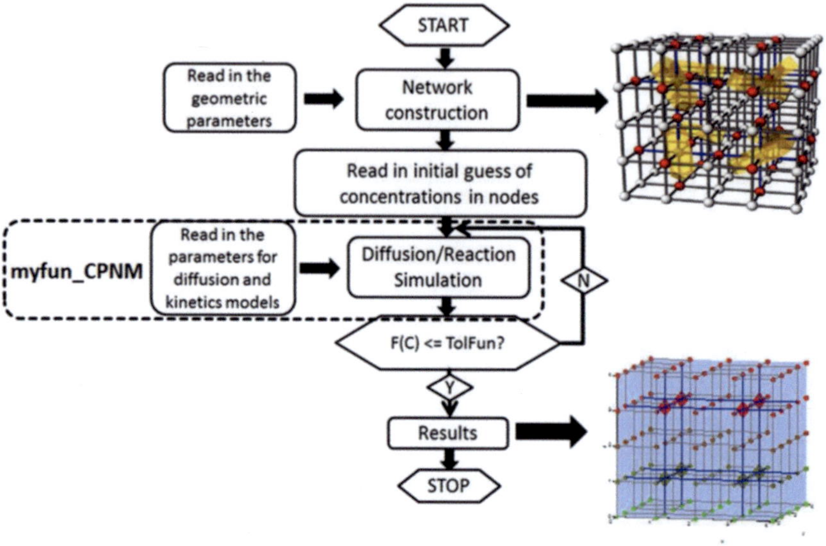

Figure 3-4: The overall algorithm of the CPNM simulating diffusion and reaction in polycrystalline microporous media.

In Step 1, the geometric parameters (e.g. network size, connection probability of intercrystalline pores, orientational angles) were read in to construct the crystallite-pore network. In the construction of a crystallite-pore network of $N = L{\times}L{\times}H$ nodes (i.e., $\frac{L-1}{2}\times\frac{L-1}{2}\times\frac{H-1}{2}$ crystallite[3]), a 4 by N matrix $\mathbf{A}$ is created. As shown in Table 3-1, the coordinate values of the nodes in the crystallite-pore network, i.e., x-, y-, z-value, will be saved in the first three rows of this matrix, respectively, while the values from 1-4 are saved in the fourth row representing the type of node. The values 1-4 mean that the node is the intercrystalline node not connected with the interface node, the intercrystalline node connected to the interface node, the interface node between two adjacent crystallite nodes, and the node representing a crystallite, respectively. The x-, y-, z-value of the nth node among N nodes can be calculated with the equations in Table 3-1, while its type-value is given according to the description of the crystallite-pore network in Section 3.1. In the network construction during the computation, this matrix can help to find the adjacent nodes and connect them more efficiently. Furthermore, it can help to define if a pore is a micropore or not.

In Step 2, the large nonlinear equation system in Eq. 3—11 written in the function file is solved with the function 'fsolve' from the Optimization Toolbox® in Matlab®. Its default algorithm – "trust-region dogleg" was chosen. In this algorithm, the trust region sub-problem, which decides whether the new result $\mathbf{c}_{k+1}$ is better or worse than $\mathbf{c}_k$, was solved *via* the dogleg approach. For a detailed mathematical description, see Powell [55]. In the iteration process, the parameters for diffusion and reaction were read in to calculate the flows into the nodes of the network and the reactions in the nodes representing the crystallites. In the initial guess for the iteration process, the concentration distribution across the network was assumed to be linear. During every iteration step, the nonlinear equation system $\mathbf{F}(\mathbf{c})$ was calculated using the diffusion and kinetic models introduced in the last sub-section. An improved guess was obtained by the trust region algorithm, while the trust region sub-problem was solved by the dogleg approach. This process stopped until the component material balances at all nodes were fulfilled, i.e., the function value of $\frac{1}{2}\|\mathbf{F}(\mathbf{c})\|^2$ was smaller than the termination tolerance.

In Step 3, in order to ensure the results to be correct, the component material balances at all nodes are checked, i.e., whether every element in the final $\mathbf{F}(\mathbf{c})$ is much smaller than its corresponding flows and/or reaction terms. Finally, using the obtained concentrations of species at every node, the fluxes of the species through the CPNM, which represents a polycrystalline membrane, can be calculated. If reactions take place in the membrane, conversion of feed gas and selectivities towards products can be predicted. Finally, the species concentration in the pore network can be displayed in Step 4 if necessary.

Table 3-1: Representation of elements in matrices in the simulation code for the crystallite-pore network model.

Matrices A = [x, y, z, type]	**Representation**
x	coordinate value in x-direction, x = *mod(n, L);
y	coordinate value in y-direction, y = §fix(mod(n, L^2)/L);
z	coordinate value in z-direction, z = fix(n/(L^2));
type	1. intercrystalline node not connected to the interface node; 2. intercrystalline node connected to the interface node; 3. interface node; 4: node representing a crystallite.

* mod function: modulus after division; §fix function: round towards zero.

4 Experimental determination of adsorption and diffusion parameters for simulation

In the simulation models mentioned in Chapter 3, SDM was used to simulate the transport of gas molecules in the intracrystalline micropores of ZSM-5 catalysts, e.g., CO, which have a molecular size much smaller than the pore size, while SFDM served to model the transport of large gas molecules close to the pore size, e.g., xylene. Both SDM and SFDM consider that the gas molecules diffuse in adsorbed state on the pore wall. Therefore, the diffusional fluxes of gases in the micropores depend on the amount of adsorbed molecules and their mobility on the pore wall. In order to simulate the diffusion in micropores, both reliable adsorption and diffusion coefficients are required in SDM and SFDM.

For the simulations, these parameters should be determined experimentally, or obtained from the literature, if available. Here, a device based on a high-temperature high frequency oscillating microbalance was established for determination of the adsorption parameters in ZSM-5 close to the reaction temperatures, while a high-pressure and high-temperature Wicke-Kallenbach cell was set up to measure the permeation of gases through the ZSM-5 membrane for determination of the corrected surface diffusion coefficients. Both setups were used for the determination of adsorption parameters and corrected surface diffusion coefficients of related gases in the direct DME synthesis reaction from syngas in ZSM-5, which is the shell catalyst of the core-shell catalyst for direct DME synthesis.

4.1 Langatate crystal microbalance adsorption device

Conventional devices for gas adsorption investigation in zeolites are based on volumetric or gravimetric methods [56]-[59]. In recent years, more and more attention has been paid to a new technique using high frequency oscillating microbalances, e.g., a quartz crystal microbalance (QCM) [60]-[61], since it is easy to operate, has lower cost and higher measurement accuracy [62]. Compared to QCM which is limited to temperatures below 80 °C [60], the langatate crystal microbalance (LCM) in principle can be used at higher temperatures, i.e., at or close to the reaction temperatures of one-stage DME synthesis from syngas, owing to the absence of crystalline phase transitions up to its melting point (1470 °C) [63]. Moreover, langatate has a piezoelectric coefficient up to 3 times higher than quartz [63]. Thus, a commercial LCM was chosen here as the acoustic wave substrate for the high-temperature gas adsorption device to study the gas adsorption in H-ZSM-5. It was used to measure the single component adsorption isotherms of CO_2, H_2O, methanol, and DME in H-ZSM-5 at temperatures between 50 and 150 °C and pressures up to 18 bar. From these adsorption isotherms, the adsorption parameters such as adsorption capacities, enthalpies, entropies were determined *via* fitting with the single or dual site Langmuir model. The results could already be published in the Journal of Physical Chemistry C [64]. Moreover, the details about the zeolite synthesis on LCM *via* SAC, as well as the adsorption measurments using the LCM device, will be published in an open-access video article in the Journal of Visualized Experiments [65].

4.1.1 Zeolite coating onto the LCM-sensor

Prior to the adsorption measurements, the zeolite ZSM-5 was coated on two LCM sensors in order to oscillate with this acoustic wave substrate, together with the colleagues at CRT. One of the two coated LCM sensors was solely used for analyzing the zeolite coating, while the other one was used to perform the adsorption experiments (see following paragraphs). In this section, the detailed coating procedure is introduced.

Figure 4-1 shows the LCM sensor employed in this study, which was purchased from *C3 Prozess- und Analysentechnik GmbH*, Munich, Germany. It has a diameter of 14 mm, a resonant frequency of 5 MHz, and polished gold electrodes on both sides. Since the zeolite coating on top of the connection points (see Figure 4-1) of the gold electrodes with the oscillator would significantly reduce the electroconductivity and thus the measurement sensitivity of the LCM, special attention was paid to solely coat in the centor of the sensor without covering the connection points (see marked region in Figure 4-1). Therefore, by the method of steam-assisted crystallization (SAC) two LCM sensors were coated in parallel, adopting the preparation procedure reported by Iglesia et al. (2007) [66]. The zeolite synthesis mixture was prepared using tetraethyl orthosilicate (Alfa Aesar, > 99 %), aluminium nitrate nonahydrate (Chempur, > 98.5 %) and tetrapropylammonium hydroxide (Sigma-Aldrich, 1 mol dm^{-3} aqueous solution) as silicon source, aluminum source and template, respectively. Alkalinity was further increased using sodium hydroxide (Merck, purity > 99 %).

The final synthesis mixture had the following molar composition: $1\ SiO_2 : 47\ H_2O : 0.07\ Na_2O : 0.12\ TPA_2O : 0.005\ Al_2O_3$, as reported by Iglesia et al. (2007) [66] and was carefully dropped on the back side of the LCM as indicated in Figure 4-1 using a pipette. *Via* subsequent drying at 80 °C for 2 hours a highly viscous gel-like phase was obtained, covering the intended region of the sensor. The coated LCM-sensor was placed horizontally in the synthesis autoclave (80 mL, Teflon lined) onto a Teflon spacer to ensure separation of the sensor from the liquid water (10 mL) at the bottom of the autoclave. SAC was performed at 150 °C for 48 hours. Right after SAC, the sensor was washed with de-ionized water and dried at 80 °C for 2 hours. Finally, the organic template was removed by calcination under oxidative atmosphere at 450 °C for 4 hours with a temperature increasing and decreasing rate of 3 K min^{-1}. The Na-ZSM-5 crystals on the LCM sensor were ion-exchanged 2 times at 20 °C for 2 hours using a 1 mol dm^{-3}

NH_4Cl solution to get the NH_4-ZSM-5. The H-form of the zeolite was obtained by final calcination using the same parameters as above mentioned.

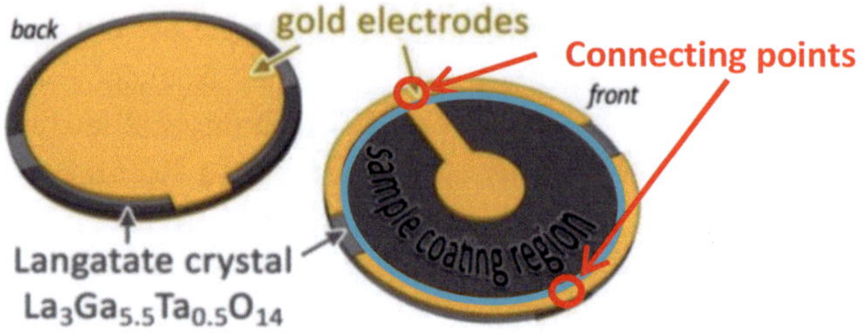

Figure 4-1: Scheme of langatate crystal microbalance (LCM) sensor in front and back side view, adopted from [64].

4.1.2 Characterization of the zeolite coating

The characterization of the zeolite coating *via* XRD, wavelength-dispersive X-ray spectroscopy (WDX), SEM and light microscopy was performed by Mr. Klumpp, a PhD student at CRT at Erlangen-Nürnberg University, and Mr. Messerschmidt at IMVT. To qualitatively prove the successful synthesis of zeolite ZSM-5 (MFI topology) on the LCM-sensors, X-ray diffraction patterns were recorded using a PANanalytical Empyreon diffractometer. As the total amount of zeolite on the LCM-sensor is very small (about 0.5 mg, see Table A1 in Appendix a)), extended X-ray exposure time was applied in the Bragg angle regime of 22 – 28 °2Θ where characteristic MFI peaks were to be expected (MFI reference pattern from ICDD, powder diffraction file no. 01-079-1638). To check for the Si/Al molar ratio of the loaded ZSM-5 crystals, an electron beam microprobe (JEOL JXA-8530F) was used to obtain WDX data of 10 randomly chosen points on the loaded ZSM-5 crystals. Moreover, to check for the morphology, dimensions and distribution of the zeolite crystals on top of the LCM-sensor, light microscopy (Nikon Eclipse 50i equipped with Nikon DS-Fi1 camera) and scanning electron microscopy (SEM, Carl Zeiss Ultra55, 2 keV) were used.

As can be seen by bare eyes (see also photographs in Figure 4-3a), a coating was achieved throughout the complete langatate crystal. Comparison of the x-ray diffraction patterns of an uncoated sensor and the coated one clearly proofs the formation of an MFI-type zeolite: Besides the highly intensive reflexes originating from the langatate crystal, typical MFI-reflexes are additionally present in the pattern of the coated sample (see inset in Figure 4-2). The obtained WDX data of 10 randomly chosen points on the loaded ZSM-5 crystals confirm that the ZSM-5 crystals had a Si/Al molar ratio of 100 reflecting the composition of the initial zeolite synthesis mixture.

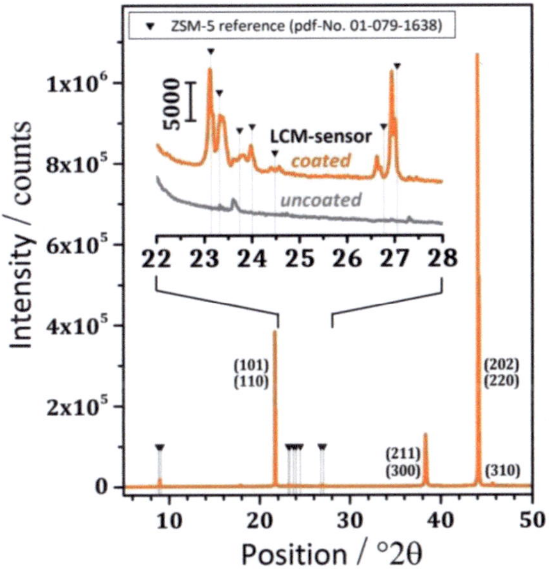

Figure 4-2: X-ray diffraction patterns of the uncoated and coated LCM-sensor. Patterns of the inset were determined applying extra long exposure times for the sake of sufficient intensity, adopted from [64].

From both, light and scanning electron microscopy (Figure 4-3b and c), the region of the gold electrode was shown to be less covered by zeolite crystals than the exposed langatate crystal region. Most of the crystals on top of the

LCM-sensor are isolated and show characteristic rounded-boat morphology, with the (010)-plane predominantly facing up. Besides, some crystals additionally show the typical intergrowth behavior ("twinned crystals"). Size distribution of the crystals is quite narrow: based on SEM image analysis a mean length of $L_a = 21.44 \pm 1.07$ µm and $L_c = 26.86 \pm 1.23$ µm in the crystallographic *a*- and *c*-direction, respectively were determined. These values fit well to the dimensions $L_a = 20.14 \pm 3.30$ µm and $L_c = 25.71 \pm 3.30$ µm obtained from light microscopy image analysis.

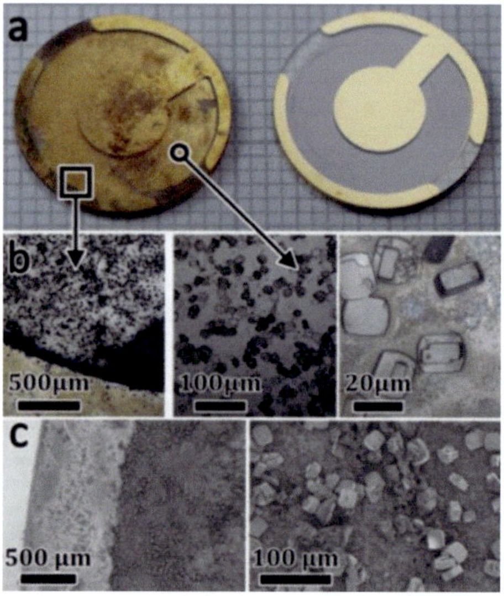

Figure 4-3: Coated langatate crystal microbalance sensor. (a) Photographs of the coated (left) and uncoated sensor (right), (b) light microscopy and (c) scanning electron microscopy images, adopted from [64].

4.1.3 Design of the high-temperature adsorption device based on a LCM

The coated LCM was used in the LCM adsorption device to determine the adsorption isotherms at high temperature. A schematic of the LCM adsorption device is shown in Figure 4-4, a labeled photograph can be found in Appendix a). In this device, a custom-designed crystal holder inside a 300 mL high pressure stainless steel chamber (purchased from *Büchi AG*, Uster, Switzerland) was used. With the given setup, two crystal sensors at the same time can be mounted, thus enabling to analyze the langatate crystal coated with the H-ZSM-5 sample and an unloaded langatate crystal reference sample in parallel. The langatate crystals were connected to an oscillator in a commercial frequency oscillating microbalance (eQCM 10M™) purchased from *Gamry Instruments*, Warminster, USA, *via* high-temperature resistant cables. The output frequencies and damping voltages of both crystals were measured by a frequency counter integrated in the eQCM 10M™ device. The resonant frequencies were determined *via* fitting and recorded by the supporting software for the eQCM 10M™ device - Gamry's Resonator™ (for details, see Figure A2 in Appendix a)).

Since preliminary work had shown that precise temperature control of the crystals is crucial to ensure a high measurement accuracy of the LCM, a thermocouple close to the crystals was used to monitor the temperature of the crystals inside the chamber. Moreover, the sample chamber placed inside the temperature control heating mantle was further insulated using glass wool, enabling the temperature inside the chamber, which was controlled by the temperature controller ("btc 01", temperature range –100/550 °C, resolution 0.02 K, *Büchi AG*), to be kept with a precision of ± 0.1 K at 50 °C, ± 0.5 K at 100 °C, and ± 1 K at 150 °C. In order to precisely monitor the absolute pressure inside the sample chamber, both high- and low-pressure transducers (0-50 bar and 0-1 bar) were applied. The sample chamber can be operated up to 275 °C temperature and 300 bar pressure. The temperature limit is determined by the maximum operating temperature recommended for the O-ring (Kalrez 6375), which is used to seal the chamber at high pressure.

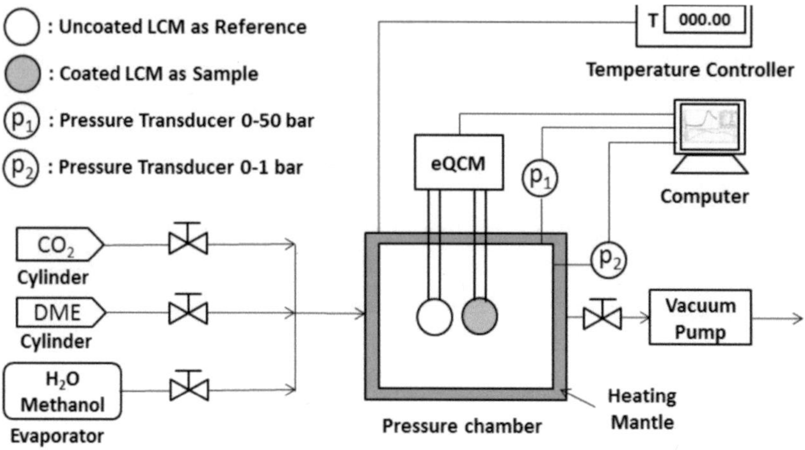

Figure 4-4: Schematic of the LCM-based high-temperature high-pressure adsorption measurement apparatus, adopted from [64].

4.1.4 Adsorption measurements

After cleaning the O-ring, the crystal holder and the sample chamber, the reference and sample crystals were placed in the holder. Before the adsorption measurements, the sample crystal loaded with H-ZSM-5 was vacuum-activated at 200 °C overnight, i.e., only negligible gas amount was adsorbed on the H-ZSM-5. During the adsorption measurement, the pressure inside the sample chamber was controlled by dosing pure gas from gas cylinders *via* mass flow controllers or from the evaporator manually *via* a dosing valve into the chamber. When equilibrium conditions and stable temperature were reached, the resonant frequencies of the reference and sample crystals were measured and recorded by the supporting software - Gamry's Resonator™ (see Figure A2 in Appendix a)). Adsorption equilibrium was defined as being reached as soon as constant temperature was observed, i.e., the measured temperature stayed within the controlling precision of the temperature controller mentioned above. Considering the fact, that by nature, the two bare langatate crystals are not identical, the frequency change of the sample crystal Δf was calculated by the following equation:

$$\Delta f = f_{0,s} - f_s = f_{0,r} - \Delta f_0 - f_s, \qquad \text{Eq. 4—1}$$

where $f_{0,s}$ and $f_{0,r}$ are the resonant frequencies of the reference and unloaded sample langatate crystals in Hz, respectively, $\Delta f_0 = f_{0,r} - f_{0,s}$ is the difference of the resonant frequencies between the reference and unloaded sample langatate crystal in Hz, which was measured before synthesizing the H-ZSM-5 crystals on the langatate crystal and shows a significant dependence on temperature but not on pressure, f_s is the resonant frequency of the langatate crystal loaded with H-ZSM-5 in Hz. The frequency change Δf is the sum of five independent terms, i.e., Δf_m, Δf_T, Δf_p, Δf_V, and Δf_R due to changes in mass, temperature, pressure, viscosity and roughness loading, respectively [61]:

$$\Delta f = \Delta f_m + \Delta f_T + \Delta f_p + \Delta f_V + \Delta f_R. \qquad \text{Eq. 4—2}$$

Since both, reference and sample langatate crystals were placed in the same chamber and the equilibrium condition was reached, the frequency changes due to temperature, pressure, viscosity, and "roughness loading" (i.e., shear caused by the ambient medium) can be neglected, i.e., $\Delta f \approx \Delta f_m$.

According to the Sauerbrey equation [62], the mass of H-ZSM-5 deposited on the sample langatate crystal and gas adsorbed on the H-ZSM-5 can be determined:

$$\Delta m = \frac{n \cdot \Delta f_m \cdot \sqrt{\rho_L \cdot \mu_L}}{2 f_{0,s}^{\ 2}} \cdot A = \frac{n \cdot (f_{0,s} - f_s) \cdot \sqrt{\rho_L \cdot \mu_L}}{2 f_{0,s}^{\ 2}} \cdot A, \qquad \text{Eq. 4—3}$$

where Δm is the difference in mass per unit area in g cm^{-2}, ρ_L is the density of the langatate crystal (6.13 g cm^{-3}) [63], μ_L is the effective piezoelectrically stiffened shear modulus of the langatate crystal (1.9×10^{12} g cm^{-1} s^{-2}) [63], n is the number of the harmonic at which the crystal is driven (in this study, $n=3$), and A is the area of the LCM-sensor in cm^2. The Sauerbrey equation is based on the important assumption that the additional mass or coating deposited on the high frequency oscillating crystal has the same acoustic-elastic properties as the crystal itself [62]. This assumption is valid if the frequency change in the measurement is small, i.e., $\Delta f_m / f_{0,s} < 1\%$ [61]-[62]. This was strictly true in all the measurements of the present study. Moreover, the sensitivity of the LCM used in this work ($C_f = \Delta f_m / \Delta m$) is about 4.4×10^7 Hz cm^2 g^{-1}, i.e., the frequency change of 1 Hz represents a change in mass per unit area of 2×10^{-8} g cm^{-2}.

As mentioned, the LCM device was used to measure the adsorption isotherms of CO_2, H_2O, methanol, and dimethyl ether in H-ZSM-5 in the temperature range of 50-150 °C and pressure range of 0-18 bar. In Table A1 in Appendix a), the detailed experimental data and calculated values for adsorption of CO_2 in H-ZSM-5 (T = 50 °C, p = 0-15.5 bar) are shown to give an example. The adsorption isotherms were fitted with a single and a dual site Langmuir model described as follows:

Singe site Langmuir: $q = q_{s,1}\dfrac{K_{s,1}p}{1+K_{s,1}p}$, Eq. 4—4

Dual site Langmuir:

$$q = q_{s,1}\frac{K_{s,1}p}{1+K_{s,1}p} + q_{s,2}\frac{K_{s,2}p}{1+K_{s,2}p}, \quad \text{Eq. 4—5}$$

where q is the gas loading in H-ZSM-5 in mmol g^{-1}, q_s is the gas adsorption capacity in the H-ZSM-5 crystals in mmol g^{-1} and K is the adsorption equilibrium constant in bar^{-1}. The temperature dependency of the adsorption equilibrium constant K can be described by the usual van 't Hoff relation:

$$K = \exp\left(\frac{-\Delta H_{ads}}{RT} + \frac{\Delta S_{ads}}{R}\right), \qquad \text{Eq. 4—6}$$

where R is the universal gas constant in J mol^{-1} K^{-1}, ΔH_{ads} represents the adsorption enthalpy in J mol^{-1} and ΔS_{ads} represents the adsorption entropy in J mol^{-1} K^{-1}.

4.1.5 Results of adsorption measurements

4.1.5.1 CO_2 adsorption

CO_2 adsorption isotherms in the temperature range of 50-150 °C and pressure range of 0-16 bar for the H-ZSM-5 zeolite are shown in Figure 4-5. The experimental data were fitted with the single-site Langmuir model. The results of the model fitting, i.e., the fitting lines of adsorption isotherms, the determined adsorption parameters, and the temperature dependency of the adsorption constant are presented in Figure 4-5, Table 4-1 and Figure A3 in Appendix a), respectively. The high quality of the fit of the single site Langmuir model for CO_2 supports the assumption of a constant adsorption enthalpy (i.e., heat of adsorption) to be valid at least for the range of conditions used. The evaluated adsorption capacity, adsorption enthalpy and adsorption entropy for CO_2 in H-ZSM-5 are 4.0±0.2 mmol g^{-1}, 15.3±0.5 kJ mol^{-1} and 56.3±1.5 J mol^{-1} K^{-1}, respectively. They compare well to values reported in literature [67]-[70], i.e., the adsorption capacity, adsorption enthalpy and adsorption entropy reported for CO_2 in MFI-type zeolites vary in the range of 2.1-3.8 mmol g^{-1}, 19-28.7 kJ mol^{-1}, and 43.7-82.7 J mol^{-1} K^{-1}, respectively, in the temperature range of 30-200 °C and pressure range of 0-5 bar (see Table 4-1).

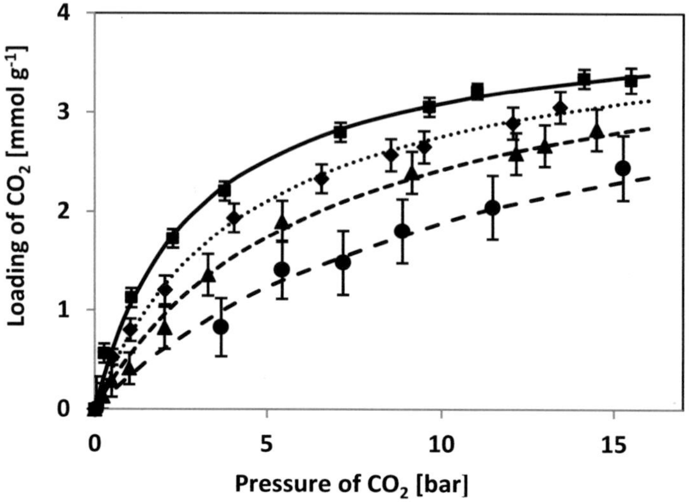

Figure 4-5: Adsorption isotherms for CO_2 in H-ZSM-5 at 50 (■), 75 (◆), 100 (▲), and 150 °C (●). The symbols represent the experimental data, the error bars indicate the measurement uncertainty, and the lines represent the fit of the single site Langmuir model to the experimental data, adopted from [64].

4.1.5.2 Water adsorption

Water adsorption isotherms for the H-ZSM-5 zeolite at 50, 85 and 120 °C were measured in the pressure range of 0-0.56 bar and are reported in Figure 4-6. The single and dual site Langmuir models were used to fit the experimental data. Their standard errors of the fitting to the experimental data are shown in Table 4-2. It can be clearly seen that fitting with the dual site Langmuir model has significantly smaller standard errors than fitting with the single site Langmuir model. Therefore, only the fitting results of the dual site Langmuir model are shown in Figure 4-6, Table 4-3 and Figure A3 in Appendix a).

Table 4-1: Adsorption parameters for CO_2, from a fit of the data of the single site Langmuir model and comparison with literature, adopted from [64].

MFI-type zeolites (Si/Al)	conditions	q_s mmol g^{-1}	$-\Delta H_{ads}$ kJ mol^{-1}	$-\Delta S_{ads}$ J mol^{-1} K^{-1}	method	ref.
H-ZSM-5(100)	50-150 °C 0-16 bar	4.0±0.2	15.3±0.5	56.3±1.5	LCM	This work
H-ZSM-5(27)	50-200 °C 0-1 bar	2.1	28.7	82.7	TPD	[67]
H-ZSM-5(31) Silicalite-1(∞)	30-200 °C		26.1 21.7		GCP	[68]
Silicalite-1(∞)	30-80 °C 0-1.8 bar	3.8	19-21	43.7-63.7	gravimetric	[69]
Silicalite-1(∞)	30-135 °C 0-5 bar	2.97	25	81.4	TEOM	[70]

As shown in Table 4-3, according to the dual site Langmuir model the adsorption capacity, adsorption enthalpy and entropy for water on site 1 are 3.9±0.5 mmol g^{-1}, 77.2±1.0 kJ mol^{-1} and 217.0±2.0 J mol^{-1} K^{-1}, respectively, while on site 2 these values are 0.4±0.1 mmol g^{-1}, 69.2±2.0 kJ mol^{-1} and 144.7±3.0 J mol^{-1} K^{-1}, respectively. These determined adsorption parameters for water in the H-ZSM-5 zeolite compare again well to values for the MFI-type zeolites reported in literature [71]-[73], which were measured at temperatures of 30 to 200 °C and pressures of 0 to 1 bar (see Table 4-3).

From the adsorption enthalpy and entropy, the Langmuir adsorption constant on site 2 (strong adsorption sites, e.g., $K_{s,2}$ (85 °C) = 400 bar^{-1}) was found to be higher than on site 1 (weak adsorption sites, e.g., $K_{s,1}$ (85 °C) = 1.1 bar^{-1}) which is in line with what can be expected from literature where the existence of weak and strong adsorption sites for water in ZSM-5 zeolites is described [73]-[76]: strong adsorption sites are thereafter related to strong Brønsted acid sites in the ZSM-5 crystallites originating from bridging silanol groups (≡Si – OH – Al≡), while terminal silanol groups (≡Si – OH) form the weak adsorption sites. Thus, the amounts of strong and weak adsorption sites correlate with the amount of framework alumina [76] and the crystal size and number of defects [71], [76], respectively. Accordingly, in the zeolite sample of the present study, the number of Brønsted acid sites should be quite low (Si/Al=100) for the reason of which the accordant adsorption capacity $q_{s,2}$ was found to be lower than $q_{s,1}$ attributed to the terminal silanol groups.

4.1.5.3 Methanol adsorption

Adsorption isotherms for methanol in the H-ZSM-5 zeolite (T = 50-150 °C, p = 0-5 bar) are shown in Figure 4-7. Since the experimental data at 150 °C have significant measurement uncertainty of 0.25 mmol g^{-1} compared to the loadings of methanol (< 0.8 mmol g^{-1}), only the experimental data at 50, 75 and 100 °C were used in the fit by the single site Langmuir model, while the experimental data at 150 °C were used only for verification of the fitted adsorption parameters. The results of the model fitting are shown in Figure 4-7, Table 4-4 and Figure A3 in Appendix a). The high fitting quality of the single site model supports the assumption that a constant heat of adsorption for methanol is valid in the temperature range of 50-100 °C.

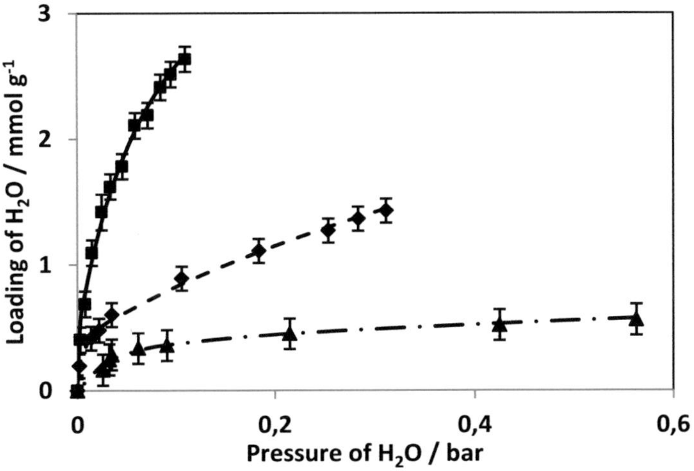

Figure 4-6: Adsorption isotherms for water in H-ZSM-5 at 50 (■), 85 (◆), and 120 °C (▲). The symbols represent the experimental data, the error bars indicate the measurement uncertainty and the lines represent the dual site Langmuir model fitted to the experimental data, adopted from [64].

Table 4-2: Standard errors of fitting with the single and dual site Langmuir models to the data of water adsorption in H-ZSM-5, adopted from [64].

Temperatures / °C	**Standard errors / mmol g^{-1}**	
	Single site Langmuir	**Dual site Langmuir**
50	0.11	0.06
85	0.10	0.025
120	0.027	0.023

Table 4-3: Adsorption parameters for water, from a fit of the data to the dual site Langmuir model and comparison with literature, adopted from [64].

MFI-type zeolites (Si/Al)	conditions	q_s mmol g^{-1}	$-\Delta H_{ads}$ kJ mol^{-1}	$-\Delta S_{ads}$ J mol^{-1} K^{-1}	method	ref.
H-ZSM-5(100)	50-120 °C 0-0.56 bar	3.9±0.5 (site 1) 0.4±0.1 (site 2)	77.2±1.0 (site 1) 69.2±2.0 (site 2)	217.0±2.0 (site 1) 144.7±3.0 (site 2)	LCM	This work
H-ZSM-5(37.5) H-ZSM-5(103)	50-200 °C 0-1 bar		42-84 42-75		thermos-gravimetric	[71]
H-ZSM-5(3.8) Silicalite-1(∞)	30-200 °C 0-0.01 bar	4	60-80 40-60		calorimetric & volumetric	[72]
Na-ZSM-5 (130)	30-200 °C 0-0.02 bar	5.34 (site 1) 0.5 (site 2)	72 (site 1) 58 (site 2)	204.3 (site 1) 110.3 (site 2)	combined IR & gravimetric analysis (AGIR)	[69]

The results displayed in Table 4-4 show that the adsorption capacity, enthalpy and entropy for methanol in H-ZSM-5 zeolite are 3.5±0.2 mmol g^{-1}, 26.3±0.3 kJ mol^{-1} and 86.4±0.8 J mol^{-1} K^{-1}, respectively. In literature [77]-[79] it is reported that the adsorption capacity, enthalpy and entropy for methanol in MFI-type zeolites in the temperature range of 20 to 55 °C vary in the range of 2-3.5 mmol g^{-1}, 35-85 kJ mol^{-1}, 80-140 J mol^{-1} K^{-1}, respectively, (see Table 4-4). Apparently, the adsorption parameters determined by LCM compare well to these literature values. Moreover, Figure 4-7 shows that the predicted adsorption isotherm for 150 °C based on the determined adsorption parameters can reproduce the experimental data within the limits of uncertainty.

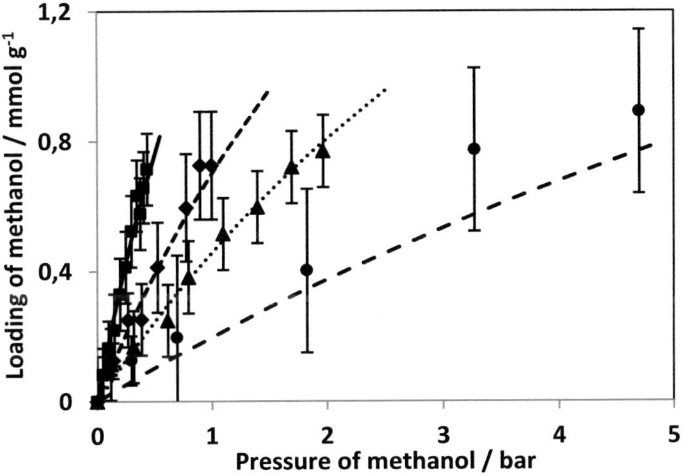

Figure 4-7: Adsorption isotherms for methanol in H-ZSM-5 at 50 (■), 75 (◆), 100 (▲), and 150 °C (●). The symbols represent the experimental data, the error bars indicate the measurement uncertainty, data at 150 °C were not used for the fit but for validation, and the lines represent the fit of the single site Langmuir model to the experimental data, adopted from [64].

Table 4-4: Adsorption parameters for methanol, from a fit of the data of the single site Langmuir model and comparison with literature, adopted from [64].

MFI-type zeolites (Si/Al)	conditions	q_s mmol g^{-1}	$-\Delta H_{ads}$ kJ mol^{-1}	$-\Delta S_{ads}$ J mol^{-1} K^{-1}	method	ref.
H-ZSM-5(100)	50-100 °C 0-5 bar	3.5±0.2	26.3±0.3	86.4±0.8	LCM	This work
H-ZSM-5(15)	25 °C 0-0.16 bar	3.5	50-85		TPD	[71]
H-ZSM-5(990)	20-55 °C 0-0.04 bar	2-3.5	44		gravimetric	[72]
H-ZSM-5(36) Silicalite-1(∞)	50 °C		47-74 35-60	110-130 80-140	calorimetric	[69]

4.1.5.4 DME adsorption

Figure 4-8 presents the DME adsorption isotherms for the H-ZSM-5 zeolite at temperatures of 50 to 150 °C and pressures up to 18 bar. Again, the experimental data at 150 °C show a significantly higher measurement uncertainty of 0.38 mmol g^{-1} to be compared to the detected gas loading (< 1.2 mmol g^{-1}). Thus, only the experimental data for 50, 85 and 120 °C were used to determine the adsorption parameters *via* fitting with the single site Langmuir model, while the experimental data for 150 °C were used for verification of the determined adsorption parameters.

The results reported in Figure 4-8, Table 4-5, and Figure A3 in Appendix a) show that the single site Langmuir model fits the experimental data for 50, 85 and 120 °C well. The determined adsorption capacity, enthalpy and entropy for DME in the H-ZSM-5 zeolite are 1.7±0.2 mmol g^{-1}, 18.1±0.4 kJ mol^{-1} and 61.7±1.0 J mol^{-1} K^{-1}, respectively. Moreover, Figure 4-8 shows a good prediction of the adsorption isotherm for 150 °C using the adsorption parameters derived from the data at 50, 85 and 120 °C. Compared to the values reported by Pope [79] (40-45 kJ mol^{-1} at 50 °C, 0-2 bar), the adsorption enthalpy of DME in H-ZSM-5 determined at 50-150 °C and 0-18 bar in this work is lower by a factor of ~2 (18.1 kJ mol^{-1}).

This effect might be explained by the fact that with increasing coverage, the heat of adsorption approaches the heat of vaporization. Such an effect was also described in literature for the adsorption of water on H-ZSM-5 [71], [74]. For DME, the heat of condensation is reported [80] to be 18.5 kJ mol^{-1} (at 298.15 K) and, thus, fits quite well to the value found in the present study. On the other hand, the DME loading of $q_s = 1.7$ mmol g^{-1} corresponds to 9.8 molecules of DME per MFI unit cell (n_{DME}/n_{UC}) which is by a factor of 5 smaller than the theoretical maximum of $n_{DME}/n_{UC} = 47.6$ (assuming the specific pore volume of MFI (0.18 cm^3 g^{-1}) to be completely filled with liquid phase [74]. Pope [79] reported a more or less constant value of the heat of DME adsorption of ~ 40-45 kJ mol^{-1} for low DME loadings ranging from $3 \approx n_{DME}/n_{UC} \leq 10$ and, thus, suggesting that the loading-effect cannot be the sole origin of the deviation in the adsorption enthalpy. As the adsorption enthalpy in the present study was calculated based on the experimental values obtained at 50-120 °C, this elevated temperature might be an additional reason for the lower heat of adsorption found in the present study as compared to literature values obtained at lower temperatures.

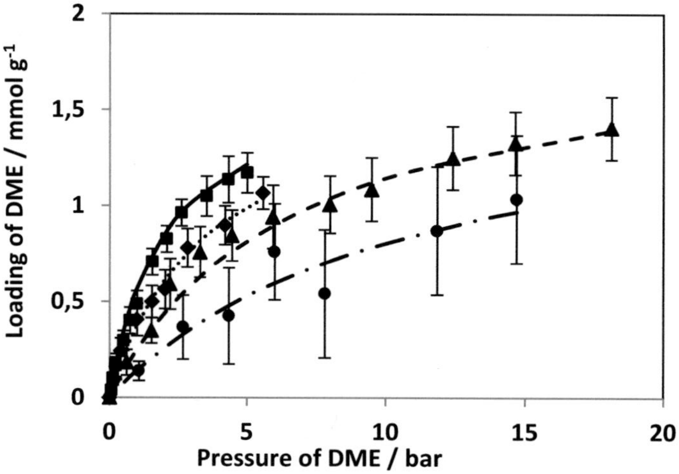

Figure 4-8: Adsorption isotherms for DME in H-ZSM-5 at 50 (■), 85 (◆), 120 (▲), and 150 °C (●). The symbols represent the experimental data, the error bars indicate the measurement uncertainty, data at 150 °C were not used for the fit but for validation, and the lines represent the fit of the single site Langmuir model to the experimental data, adopted from [64].

Table 4-5: Adsorption parameters for DME fitted from single site Langmuir model and comparison with literature, adopted from [64].

MFI-type zeolites (Si/Al)	conditions	q_s mmol g^{-1}	$-\Delta H_{ads}$ kJ mol^{-1}	$-\Delta S_{ads}$ J mol^{-1} K^{-1}	method	ref.
H-ZSM-5(100)	50-150 °C 0-18 bar	1.7±0.2	18.1±0.4	61.7±1.0	LCM	This work
H-ZSM-5(36)	50 °C		40-45	55-130	calorimetric	[69]
Silicalite-1(∞)	0-2 bar		40-45	40-100		

4.1.5.5 Comparison of adsorption isotherms at reaction temperature

As shown in Table 4-1 to Table 4-5 above, the adsorption capacities of H-ZSM-5 for the studied gases have the following order of decreasing capacity: CO_2 > H_2O (site 1) > methanol > DME > H_2O (site 2), while their adsorption enthalpies turn out as: H_2O (site 1) > H_2O (site 2) > methanol > DME > CO_2.

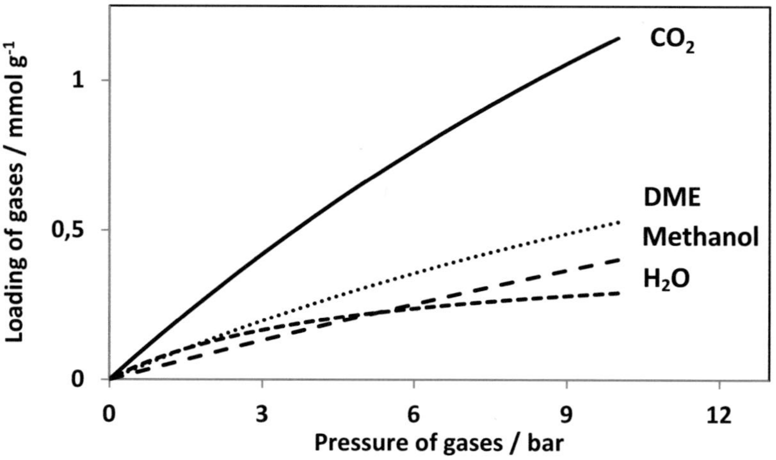

Figure 4-9: Predicted adsorption isotherms for CO_2, H_2O, methanol and DME in H-ZSM-5 at 250 °C, in the pressure range of 0-10 bar, adopted from [64].

Using these adsorption parameters, determined in the temperature range of 50-150 °C, the adsorption isotherms at 250 °C are predicted, which is a typical reaction temperature for one-stage DME synthesis. The predicted adsorption isotherms are shown in Figure 4-9. Compared to methanol, more CO_2, DME, and H_2O is adsorbed in the pressure range of 0-5 bar, which agrees with the result reported by Pope [79], whereas more methanol than

H_2O will be adsorbed at higher pressures due to the higher adsorption capacity, since site 2 is the dominant adsorption site for H_2O at this temperature.

4.2 High-pressure and high-temperature Wicke-Kallenbach cell

4.2.1 Supported ZSM-5 membranes for permeation measurements

Three ZSM-5 membranes were prepared by Ms. Reuß, another PhD candidate at CRT by multiple in-situ crystallization according to the method of Avhale [81] on disk-type porous sinter metal supports with a diameter of 18 mm. These metal supports have an excellent mechanical strength but also a high porosity and a large pore size (d_p > 250 nm) to ensure negligible transport resistance (see the cross-section SEM image of the supported ZSM-5 membrane in Figure 4-11). The supported ZSM-5 membranes were expected to be placed in a WK cell as shown in Figure 4-12 for determination of corrected surface diffusion coefficients of gases in ZSM-5.

Before sending these supported zeolite membranes to IMVT, Ms. Reuß had confirmed that they are pure ZSM-5 membranes by X-ray diffraction (XRD). At IMVT, they were also characterized with WDX and SEM by Mr. Messerschmidt at IMVT to know the composition and structure of coated ZSM-5 (e.g., Si/Al ratio, crystallite size and orientation). The overview top-view SEM image of the supported zeolite membrane (not shown here) shows that the ZSM-5 crystallites cover all the surface of the support, and form a dense membrane on the suppport. Moreover, the detailed top-view SEM image of the supported ZSM-5 membrane in Figure 4-10 shows that the most zeolite crystallites have a typical coffin-form with the size of 2 μm × 1 μm × 0.5 μm, while others are intergrown. In Figure 4-11, the cross-section SEM image of the supported ZSM-5 membrane shows that the ZSM-5 zeolite layer on the fine-porous skin layer of the support has the thickness of about 7 μm. In addition, some zeolite crystals can be found inside the support. This phenomenon was also found by Haag et al. (2006) [13] in their supported ZSM-5 membrane. An electron beam microprobe (JEOL JXA-8530F) was used to obtain WDX data of 10 randomly chosen points on the ZSM-5 crystals coated on the support. The ZSM-5 crystals are shown to have a Si/Al molar ratio of 60.

Figure 4-10: SEM top-view image of supported ZSM-5 membrane.

4.2.2 Permeation measurement apparatus based on WK cell

A Wicke-Kallenbach (WK) cell illustrated in Figure 4-12 is a simple and effective setup for studying transport and separation in porous membranes, e.g., it was used by Van de Graaf, et al. [39] to determine the corrected surface diffusion coefficients of methane, ethane and propane in silicalite-1 at pressures below 5 bar and low temperatures between 0-100 °C (silicalite-1 has the same microstructure as H-ZSM-5). In this dissertation, a high-temperature high-pressure permeation measurement apparatus based on the WK cell was established to determine the corrected surface diffusion coefficients of gases in H-ZSM-5 *via* the permeation measurements on the H-ZSM-5 membrane presented in last sub-section, as well as to know the size and concentration of defects in the H-ZSM-5 membrane.

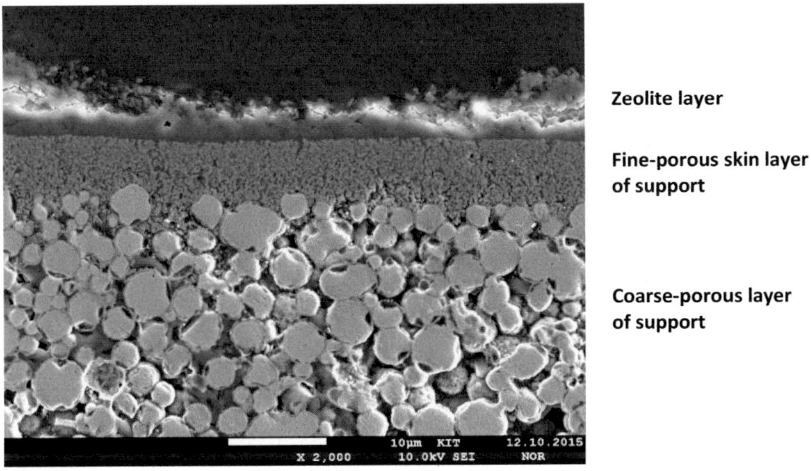

Figure 4-11: SEM cross-section image of supported ZSM-5 membrane, which consists of a coarse-porous layer of the support (dp ≈ 50 μm), a fine-porous skin layer of the support (dp ≈ 250 nm), and a zeolite layer synthesized on the skin layer.

Different from previous WK cells for low pressure and temperature, the WK cell constructed here is designed for high pressure up to 50 bar and high temperature up to 275 °C. The operating temperature is limited by the maximum operating temperature recommended for the O-ring (Kalrez 6375), which is used to seal the zeolite membranes as well as the WK cell at high pressures. More details about the dimensions and design of the WK cell can be found in Figure S4 of Appendix b).

A schematic of the permeation measurement apparatus based on the WK cell is shown in Figure 4-13, while a labeled photograph of the setup can be found in Appendix b). Prior to the permeation measurements, the supported ZSM-5 membrane is placed in the cell and sealed with the O-ring to divide the cell into two compartments - the retentate and permeate compartment. The pressure in the WK cell was controlled by two pressure transducers (PT) for lower (0-16 bar) and higher pressures (up to 50 bar), respectively, while some temperature controllers were used to achieve stable temperatures in the WK cell, as well as to control the temperatures of the tubes connected with the WK cell for the gas feed and outlet. In order to have a good temperature control, the WK cell, as well as all the connection tubes and valves in the apparatus, are isolated by surrounding the glass woll. During the permeation measurements, the exit gas compositions of the permeate and retentate chambers were measured with the gas chromatograph (GC, Agilent Technologies 7890).

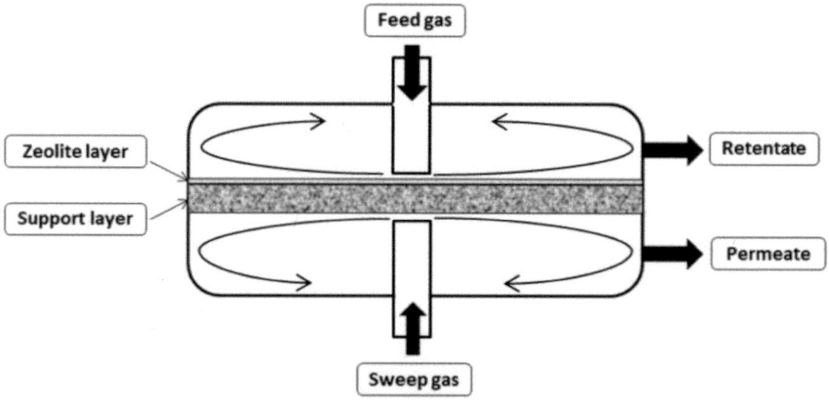

Figure 4-12: Wicke-Kallenbach permeation cell with a supported zeolite membrane, adopted from [53].

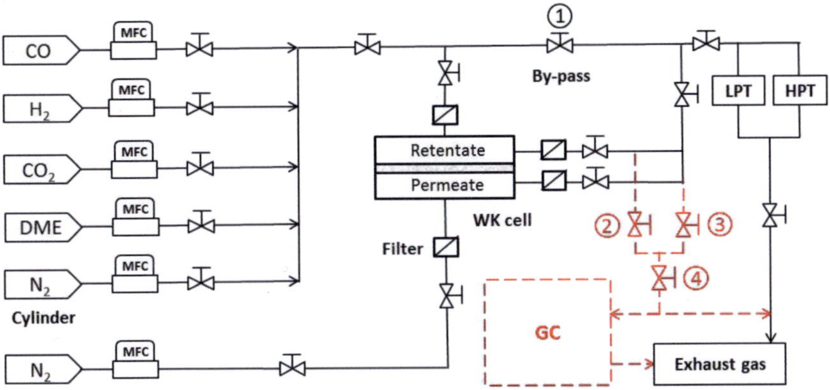

Figure 4-13: Schematic of the high-temperature high-pressure permeation measurement apparatus based on the WK cell. MFC: Mass flow controller; LPT: Low pressure transducer; HPT: High pressure transducer; GC: Gas chromatograph.

4.2.3 Tests on apparatus and sealing of zeolite membrane

4.2.3.1 Leakage test on apparatus

Since this apparatus based on the WK cell is designed for explosive and toxic gases (e.g., H_2, CO) at high pressure and high temperature, a leakage test control of all the valves and connections as well as the WK cell is fundamental porior to the permeation measurements. The test was performed with H_2 at a pressure up to 40 bar at 200 °C. The potential leakages were checked with a sensitive detector of combustible gases. Moreover, a pressure hold test at 40 bar and 200 °C showed that the pressure drop in the apparatus caused by the leakage was less than 1 bar within 10 mins. Since the total gas volume in the apparatus is about 0.3 L, the calculated leakage flow at this condition was less than 14 mL(STP)/min, which comfirmed that the apparatus had a good tightness.

4.2.3.2 Test on sealing of zeolite membrane

In order to check whether the zeolite membranes between the retentate and permeate chambers were well sealed by the O-ring at high temperature and pressure, a gas mixture (total flow: 400 mL(STP)/min) of CO_2, H_2 and N_2 with the molar composition 3.75% : 3.75% : 92.5% was fed into the retentate chamber, while the permeate chamber was fed with the pure sweep gas N_2 (total flow: 400 mL(STP)/min). The temperature and pressure in both chambers were controlled at 200 °C and 30 bar. Table 4-7 shows the gas exit or inlet flows in the retentate and permeate chambers, as well as the calculated compositions of inlet gases from the MFC data and of exit gases (i.e., well-mixed gases in the chambers) from the GC data. From the obtained gas compositions in both chambers, the permeances of CO_2 and H_2 through the ZSM-5 membrane were calculated. Their values and the membrane selectivity S_m of H_2/CO_2 derived from the permeances (i.e., permeance ratio of H_2/CO_2) are presented in Table 4-6. The results show that the ZSM-5 membrane performed a membrane selectivity of 3.72±2.12 and a H_2 permeance through the membrane was $2.54\pm0.44\times10^{-8}$ mol s^{-1} m^{-2} Pa^{-1}. The membrane selectivity of H_2/CO_2 $S_m >> 1$ and low permeances of H_2 and CO_2 indicate that the zeolite membranes between the retentate and permeate chambers are well sealed by the O-ring, even at 200 °C and 30 bar. Thus, this apparatus and the prepared zeolite membrane are ready for determination of the size and concentration (per area) of defects, and afterwards the corrected surface diffusion coefficients of gases in the ZSM-5 zeolite *via* the single gas permeation measurements.

Table 4-6: Permeances of CO_2 and H_2 through the ZSM-5 membrane, and the membrane selectivity.

Permeance of CO_2 [mol s^{-1} m^{-2} Pa^{-1}]	**Permeance of H_2 [mol s^{-1} m^{-2} Pa^{-1}]**	**Membrane selectivity [-]**
$6.83\pm2.05\times10^{-9}$	$2.54\pm0.44\times10^{-8}$	3.72±2.12

Table 4-7: Gas exit or inlet flows in the retentate and permeate chambers, the calculated compositions of inlet gases from the MFC data, and of the gas mixtures in both chambers from the GC data.

Chambers	**Feed gases**	**Inlet flows [ml(STP)/min]**	**Molar fractions in inlet flows (%)**	**Molar fractions in exit flows (%)**
Retentate	N_2	370±5	92.5±1.3	-
	CO_2	15.0±0.2	3.75±0.05	3.55±0.01
	H_2	15±0.2	3.75±0.05	3.40±0.01
Permeate	N_2	400±20	100±5	-
	CO_2	0	0	0.0620±0.0004
	H_2	0	0	0.211±0.002

4.2.4 Single gas permeation measurements for determination of corrected surface diffusion coefficients

In the permeation measurements, after closing the by-pass valve (Valve 1 in Figure 4-13), the feed and sweep gases were fed *via* the tubes close above the membrane centers on two sides. They went out of the retentate and permeate chambers *via* the tubes at the side of the WK cell. This design created vortex flow pattern in both chambers. Moreover, the concentration changes of the feed and sweep gases were small through the retentate and permeate chambers (e.g., see Table 4-7). Therefore, the gases in both chambers were assumed to be well mixed, i.e., the uniform gas concentration was achieved in both chambers of the cell. Before measuring the gas composition in the permeate chamber with the GC, Valve 2 was kept closed while Valves 3 and 4 were open for half hour, in order to sweep the impurity gases out of the apparatus. After that, a small volume of the permeate gas stream was injected into the GC to measure the gas composition in the permeate compartment. Such measurements were repeated, e.g., every 10 min, until the obtained last four GC data had a deviation smaller than 0.5 %. To measure the gas composition in the retentate chamber, Valve 2 was opened while Valve 3 was closed, and the GC measurements were repeated as introduced above.

The corrected surface diffusion coefficient of a gas in the ZSM-5 can be determined from the single gas permeation data. For such a unary system, only friction between molecules and the solid surface occurs. The equation of SDM, i.e., Eq. 2—8, is therefore simplified as follows:

$$J_i^S = -\rho q_{sat} \frac{D_i^S}{1-\bar{\theta}_i} \nabla \theta_i ,$$

$$\text{i.e., } D_i^S = -\frac{J_i^S}{\rho q_{sat} \nabla \theta_i} \cdot \left(1-\bar{\theta}_i\right). \qquad \text{Eq. 4—7}$$

where $\overline{\theta}_i$ is the average coverage of species *i* in the ZSM-5 membrane. In this equation, the coverages of gases in the H-ZSM-5 (θ_i) can be calculated with the partial pressures of gases and the adsorption parameters, e.g., obtained by the LCM adsorption measurement device as shown in the last sub-section. Thus, the corrected surface diffusion coefficients of gases in the H-ZSM-5 (D_i^S) can be obtained from the permeation flux through the zeolite pores due to the surface diffusion (J_i^S).

However, the permeation flow through the zeolite membrane measured by the WK cell consists of the flows through the zeolite pores and defect pores. Therefore, prior to the single-component permeation measurements for determination of corrected surface diffusion coefficients of gases, the size and concentration of defect pores in the H-ZSM-5 membrane should be measured for calculation of the flows in defect pores, using this established apparatus *via* e.g., n-hexane permporosimetry measurements [82]. This work is being done by the following PhD student of this DFG project at IMVT.

5 Determination of reaction parameters in core-shell catalysts

As shown in Figure 5-1, a spherical microsized core-shell catalyst ($Cu/ZnO/Al_2O_3$@H-ZSM-5) for direct DME synthesis from syngas has been synthesized by Mr. Klumpp at CRT for use in a micro packed bed reactor fabricated at IMVT. Its catalytic performance has been investigated by Mr. Lee, another PhD candidate at IMVT/IKFT[1] *via* packing in a tubular reactor, which had been established in a previous study [83]. In this previous study [83], it was packed with the admixed hybrid catalyst particles of $Cu/ZnO/Al_2O_3$ and H-ZSM-5, and the obtained experimental data were used to determine the kinetic parameters of reactions accouring in $Cu/ZnO/Al_2O_3$ and H-ZSM-5 by help of a 1-D pseudo-homogenous reactor model.

Here now, a 1-D reactor model based on a heterogeneous model for the core-shell catalyst particles was established to simulate the direct DME synthesis in the tubular reactor packed with the the prepared core-shell catalyst particles. Different from the previous reactor model in [83], it can simulate the core-shell structure, as well as the deactivation of $Cu/ZnO/Al_2O_3$ and H-ZSM-5 during the prepatation of the core-shell catalyst particles. Moreover, the heterogeneous model for the core-shell catalyst particles needs much less calculation than the complex 3-D CPNM presented in the next chapter. Therefore, it was used to determine the reaction parameters of one-stage DME synthesis in core-shell catalysts *via* fitting with the experimental data obtained by Mr. Lee [84]. In this chapter, the results of experiments and simulations on the prepared core-shell catalyst particles are shown, which have been published in [84]. Further detailed studies on the effects of the polycrystalline and anisotropic structure of the

[1] IKFT: Institute of Catalysis Research and Technology

zeolite shell layer using the more time-consuming 3-D model presented in Chapter 3 will be shown in next chapter.

5.1 Why microsized core-shell catalyst for DME production

Air pollution from fossil fuels is one of the most serious environmental problems all over the world. Pollutants (e.g., particulates, nitrogen oxide) have seriously negative effects on the environment and on human health [85]-[86]. Therefore, more and more attention has been paid in recent years to the use of oxygenated hydrocarbons (e.g. alcohols and ethers) as substitutes for fossil fuels because of their clean-burning characteristics. Dimethyl ether (DME) is such a promising clean-burning fuel for the future energy market and provides a potential alternative to diesel and liquefied petroleum gas (LPG) [87]-[88].

The production of DME from syngas is a sequential reaction system: methanol synthesis from syngas is combined with subsequent methanol dehydration to DME. In industry, its conventional production takes place *via* a two-stage process: methanol synthesis over a metal oxide catalyst (e.g., $Cu/ZnO/Al_2O_3$) in the first stage followed by methanol dehydration to DME on an acidic catalyst (e.g., zeolite H-ZSM-5) in the second stage. In recent years, a one-stage process combining both reactions in one unit using a hybrid or a bifunctional catalyst has been preferred in R&D [89], since the continuous removal of methanol *via* dehydration by the acidic catalyst improves the thermodynamic conditions of methanol synthesis and thus a higher syngas conversion can be achieved. Moreover, it was reported in [9] and [90] that when the bifunctional catalyst has a dedicated core-shell structure as shown in Figure 5-1 top (Core: methanol synthesis catalyst; Shell: methanol dehydration catalyst), the selectivity to DME can strikingly exceed the one obtained with admixed hybrid catalysts.

However, one-stage DME synthesis is strongly exothermic. To cope with this in a practical application, small-sized catalyst particles may be used in a micro packed bed reactor which enables a very high heat exchange performance (see Figure 5-1 bottom). In previous work by Tauro [83], experiments and simulations were carried out on a tubular reactor (ϕ = 8 mm, L = 0.18 m) and a micro packed bed reactor, which were both packed with a hybrid bed of commercial $Cu/ZnO/Al_2O_3$ and self-synthesized zeolite H-ZSM-5 particles with a size of d_p = 100-200 μm. The micro packed bed reactor showed better temperature control resulting in a temperature difference in the whole reactor of only 0.8 K, while the tubular reactor, despite its moderately large internal diameter of only 8 mm, already led to radial and axial temperature differences up to 3 K and 5.6 K, respectively [83]. Hence for typical industrial reactors with larger tube diameters, e.g., in the range of several centimeters, much larger nonisothermicity effects are expected.

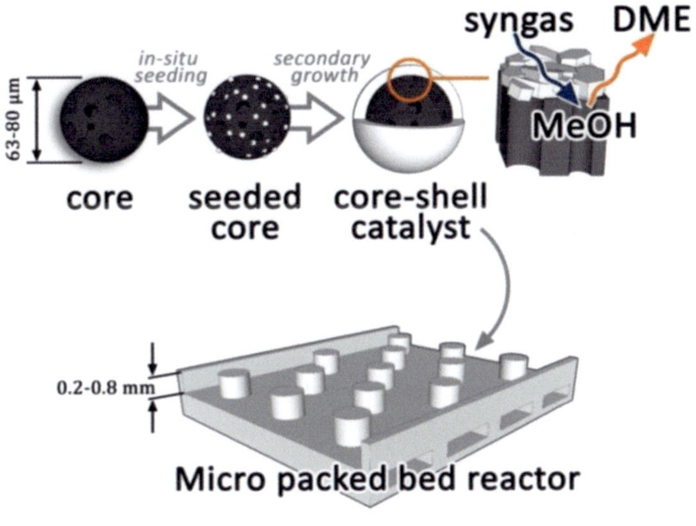

Figure 5-1: Core-shell catalyst for DME synthesis (top) packed in a micro packed bed reactor (bottom), adopted from [84].

5.2 Catalyst preparation and characterization

A microsized core-shell catalyst (Core: same commercial $Cu/ZnO/Al_2O_3$ as in [83] with d_p = 63-80 µm; Shell: zeolite H-ZSM-5) was synthesized by Mr. Klumpp at CRT by the method of *in-situ* seeding and secondary growth (see Figure 5-1) for use in a micro packed bed reactor. The details about the preparation can be found in [84].

Mr. Klumpp has proven the presence of crystalline MFI-type zeolite in the prepared core-shell catalysts by powder X-ray diffraction (XRD) patterns (See Figure 5-2). Moreover, SEM top-view and cross-section images were taken by a Carl Zeiss Ultra55 microscope (2 kV) and Jeol JXA-8530F electron microprobe (20 kV), respectively, in order to check for the macroscopic quality of the shell. The SEM top-view image of the prepared core-shell catalyst (Figure 5-3) clearly shows the successful synthesis of a very homogeneous zeolite ZSM-5 layer enwrapping the individual core particles. The shell itself is composed of randomly orientated crystals (approx. l_c = 2 µm and $l_{a,b}$ = 1 µm) which are partly intergrown.

Cross-sectional SEM analysis gives information concerning the integrity of the shell: as shown in Figure 5-4 A, the shell is completely intergrown with the core, promising good adherence and therefore high mechanical strength of the core-shell system. Based on the Si-map from EPMA/WDX analysis of the cross-section (Figure 5-4 B), a mean shell thickness of s_{shell} = 2.7 µm was determined. Using higher magnification (Figure 5-4 C), the distribution of Si, Cu and Zn within one single particle clearly underlines the above mentioned: a closed zeolite shell enwrapping the Cu/Zn-based core particle was formed with some Si also present within the core, most likely originating from the seeding step.

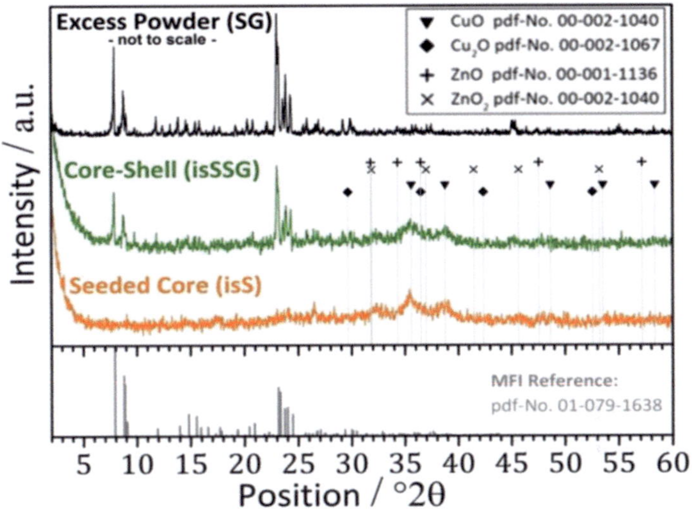

Figure 5-2: X-ray diffraction patterns of the seeded core, core-shell catalyst and zeolite excess powder, adopted from [84]. SG: in situ secondary growth; isSSG: in situ seeding and in situ secondary growth; isS: in situ seeding.

Figure 5-3: SEM top view image of core-shell catalyst, adopted from [84].

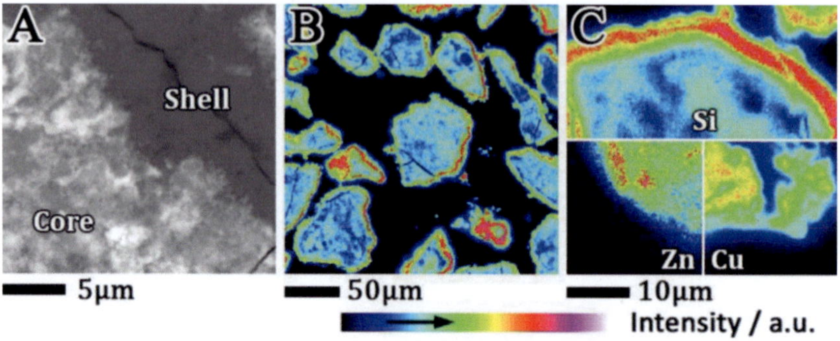

Figure 5-4: Cross-section SEM (A) and EPMA/WDX of the core-shell catalyst (B: Si-map, C: Si-, Zn- and Cu-map of one single particle), adopted from [84].

5.3 Catalytic performance investigation in a tubular reactor

In order to assist the catalyst optimization, the tubular reactor used in a previous study [83] was applied to analysis the prepared core-shell catalyst and to determine the reaction parameters for simulations. One-stage DME synthesis from syngas was carried out in this stainless steel tubular reactor (length × diameter: 18 × 0.8 cm). 6 g of the core-shell catalyst was charged to the reactor, and the rreactantion of the catalyst was performed *in situ*. For the rreactantion, the temperature was raised up to 250 °C. 3 % H_2 in N_2 was used up to 230 °C, and the temperature then was increased to 250 °C in presence of pure H_2 gas. For the reaction, syngas (H_2: CO: CO_2 = 57.6 %: 28.8 %: 3.6 %) was used and N_2 (10 %) was added as an internal standard. The reaction temperature was varied from 250 to 280 °C, whereas the reaction pressure and gas hourly space velocity were kept at 40 bar and 1000 $ml_{STP}\ g_{cat}^{-1}\ h^{-1}$, respectively. A gas chromatograph was employed for analysis of the reactant and product gases. It was equipped with HP-Plot Q and HP-Molesieve columns for separation of the gas mixtures and a flame ionization detector (FID) as well as a thermal conductivity detector (TCD) for quantification. The conversion of CO as well as selectivities to DME, methanol and CO_2 were calculated according to the following equations:

$$X_{CO} = \frac{J_{CO,feed} - J_{CO,outlet}}{J_{CO,feed}} \qquad \text{Eq. 5—1}$$

$$S_{DME} = \frac{2J_{DME,outlet}}{J_{CO,feed} - J_{CO,outlet}}, \qquad \text{Eq. 5—2}$$

$$S_{MeOH} = \frac{J_{MeOH,outlet}}{J_{CO,feed} - J_{CO,outlet}}, \qquad \text{Eq. 5—3}$$

$$S_{CO_2} = \frac{J_{CO_2,outlet} - J_{CO_2,feed}}{J_{CO,feed} - J_{CO,outlet}} = 1 - S_{DME} - S_{MeOH}. \qquad \text{Eq. 5—4}$$

As shown in Figure 5-5, the CO conversion and DME selectivity increase with the increasing reaction temperature, while the selectivity to methanol decreases with the increasing reaction temperature. The DME selectivity was at a level of around 60 %. This means the core-shell catalyst could perform the direct DME synthesis.

5.4 Models for tubular reactor with core-shell catalyst

5.4.1 Assumptions in the reactor and catalyst models

Based on a heterogeneous model for the core-shell catalyst particles, a 1-D reactor model was established to simulate the direct DME synthesis in the tubular reactor packed with the the prepared core-shell catalyst particles.

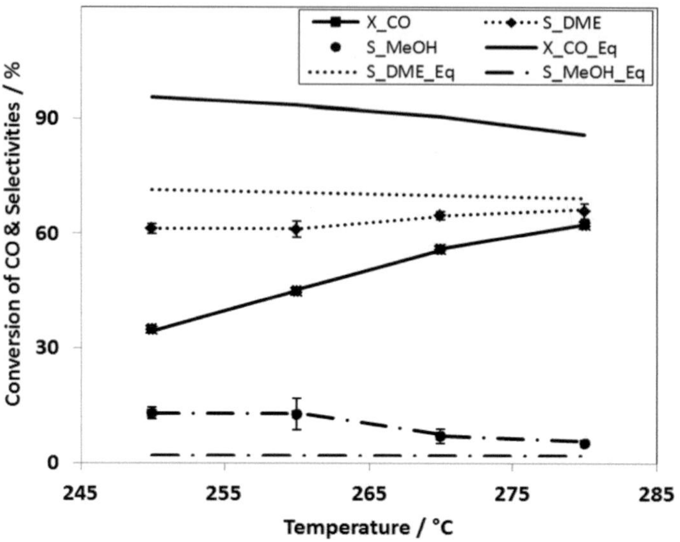

Figure 5-5: Experimental CO conversion and product selectivities in DME direct synthesis on a core-shell catalyst in a tubular reactor; Comparison with equilibrium values, at 250-280 °C. Operating conditions: 40 bar, GHSV: 1000 mlSTP/gcat/h, N_2:H_2:CO:CO_2 = 10:57.6:28.8:3.6. This diagram is adopted from [84].

Tauro [83] had evaluated the validity of the assumption of a pseudo-homogeneous system with ideal plug flow for the tubular reactor packed with admixed catalysts by experiments and applying a priori diagnostic criteria. It was shown that:

1) radial temperature gradients in the catalyst bed during the experiments were very small ($\Delta T < 3$ °C),
2) no external heat and mass transport limitation occurs around the particles based on the Anderson criterion [93] and Mears criterion [94],
3) no intraparticle temperature gradients exist based on the Anderson criterion [93], and
4) negligible axial dispersion as Bodenstein number > 100 [95].

Since the same tubular reactor and similar operating conditions were used in the investigation of the core-shell catalyst, which had smaller particle size, the following assumptions were used in the development of the 1-D reactor model based on the heterogeneous model for DME synthesis in the tubular reactor packed with the core-shell catalyst particles:

1) no radial temperature gradient in the core-shell catalyst bed,
2) no heat and mass transport limitation around catalyst particles,
3) no intraparticle temperature gradients, and
4) negligible axial dispersion.

Additionally, it was assumed that the tubular reactor was isothermal in the axial direction, since It was shown in [83] that the maximum temperature differences in the axial direction of the tubular reactor did not exceed 5.6 K even at high CO conversion of 90 %, whereas the CO conversion of the prepared core-shell catalyst shown in Figure 5-5 are lower ($X_{CO} < 60$ %).

5.4.2 1-D reactor model for the tubular reactor

In the 1-D reactor model, the reactor was discretized into 1000 sections ($n_{reactor}$) in the axial direction to simulate the concentration profiles of the species along the reactor, since the length of the catalyst bed ($L_{cat} \approx 0.1$ m) is about 1000 times of the catalyst diameter ($d_p \approx 100$ µm). Simulated species concentration profiles (see Figure 5-9) showed that the absolute differences of species concentrations between the inlet and outlet of one discretized section, i.e., $|(c_{i,out} - c_{i,in})/c_{i,in}|$ (see Figure 5-6 middle), were below 1 %. Thus, in one discretized section, the species concentrations in its bulk gas phase were assumed to be uniform and equal to the outlet species concentrations $c_{i,out}$. Moreover, for the $n_{reactor}$ consecutive sections, the total gas flow and composition in the outlet from one discretized section were used as the inlet total gas flow and composition for the next discretized section.

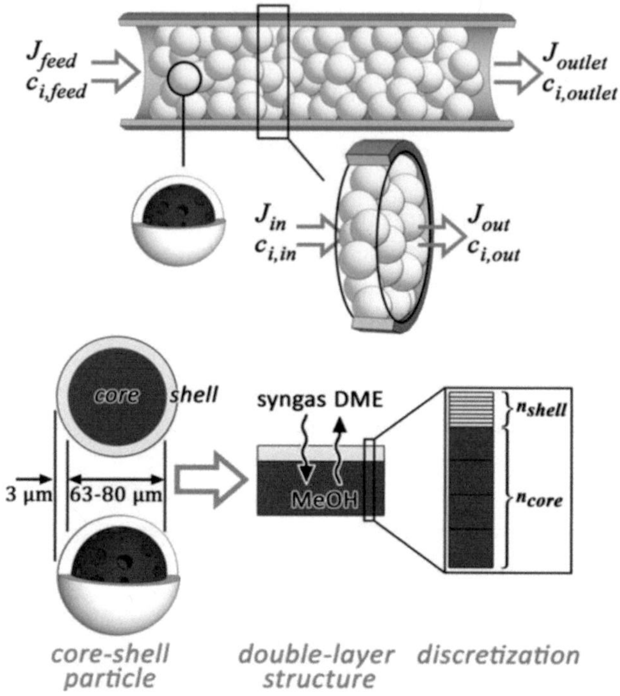

Figure 5-6: Scheme of the tubular reactor packed with core-shell catalyst (top) and one discretized section (middle). Transformation of spherical core-shell system to planar double-layer structure and discretization scheme for the double layer structure (bottom), adopted from [84].

Since the synthesis gas is converted to DME by the core-shell catalyst in every section, the gas flow and composition varies along the reactor. From the known flow and composition of the feed gas, the flow and composition of the outlet gas leaving the reactor can be obtained after $n_{reactor}$ calculations. In each calculation, an additional model, i.e., the heterogeneous model for the core-shell catalyst particles, is needed to simulate diffusion and reactions in the core-shell catalyst, which will be shown in the next sub-section.

5.4.3 Heterogeneous model for the core-shell catalyst

In the core-shell catalyst, the mass transport of the species is limited by the microporous ZSM-5 shell layer (with few defects) due to the much lower diffusivity inside the zeolite crystals. Simulation showed that discretizing the core into 4 layers, as shown in Figure 5-6 bottom (right), the absolute differences of species concentrations in core were below 1 %, i.e., negligible concentration gradients of the species exist in the core. In addition, the ZSM-5 shell layer (3 µm) is much thinner than the core catalyst (d_p = 63-80 µm). Thus, as an approximation, the spherical core-shell system can be treated as a planar double layer structure, as shown in Figure 5-6 bottom (left and middle). After the transformation, the thickness of the shell layer in the double layer structure remains unchanged (s_{shell} = 3 µm), whereas the thickness of the core layer was assumed to be $d_p/6$ µm in order to keep the same volume of the core catalyst as for the core-shell system.

As shown in Figure 5-6 bottom (right), the shell and core region in the double layer structure were discretized into n_{shell} and n_{core} layer segments, respectively, in order to account for the species concentration profiles inside the catalyst particles. The numbers n_{shell} and n_{core} were chosen as small as possible in order to reduce the computational cost, while keeping the discretization error negligible for the final simulation results. In this work, n_{shell} and n_{core} were chosen 10 and 4, respectively, according to this rule.

According to the component material balance in one discretized layer segment, it holds:

$$\nabla \mathbf{N} - m_{cat}(\mathbf{v} \cdot \mathbf{r}) = 0 \qquad \text{Eq. 5—5}$$

$\nabla \mathbf{N}$: vector of species flow gradients; m_{cat}: catalyst mass in the layer segment; $\mathbf{v}$: matrix of stoichiometric coefficients; $\mathbf{r}$: vector of reaction rates.

The Knudsen diffusion model (KDM) was used to simulate the flux of species i through the mesopores in the shell (i.e., defects, $d_p \approx 10$ nm) and in the core region $d_p \approx 25$ nm). The fluxes of species were calculated by:

$$J_{meso,i} = -\frac{\varepsilon}{\tau} \cdot D_{K,i} \cdot \nabla c_i \qquad \text{Eq. 5—6}$$

with the Knudsen diffusion coefficient of species i given by:

$$D_{K,i} = \frac{d_{p.meso}}{3} \sqrt{\frac{8RT}{\pi M_i}} \,. \qquad \text{Eq. 5—7}$$

τ : Tortuosity; ε : Porosity; $d_{p.meso}$: Diameter of mesopores. Their values used in the simulations can be found in Table 5-1. The tortuosities of mesopores and micropores in the shell were estimated by the CPNM described in Chapter 3, using the following structural parameters of the shell: connectivity of mesopores and micropores assumed to be 4, shell thickness: 3 µm, and occupied space of one crystallite: 1 µm × 1 µm × 1 µm, (see Figure 5-3). Moreover, the porosities referring to mesopores and micropores, respectively, in the shell were estimated based on the data in the previous study [53].

In this model, SDM was used to simulate diffusion in the micropores of the ZSM-5 shell layer due to their narrow pore space using the diffusivities and adsorption parameters from the literature [96]-[103] (see Table 5-1). In the calculation of the gas flows in micropores *via* SDM, the multicomponent adsorption isotherm was modeled by the extended Langmuir equations, while the mutual Maxwell-Stefan diffusivities were calculated with Eq. 2—11.

The elements of $\mathbf{N}$ are obtained from the fluxes by multiplying with the transport area A:

$$N_i^{shell} = A \cdot J_{meso,i}^{shell} + A \cdot J_{micro,i}^{shell} \,, \qquad \text{Eq. 5—8}$$

$$N_i^{core} = A \cdot J_{meso,i}^{core} \,. \qquad \text{Eq. 5—9}$$

In the model of the core-shell catalyst, since the gas flow through the catalyst bed is high, according to the Mears criterion, it is assumed that there is no transport limitation in the outside boundary layer between the catalyst surface and the bulk gas phase, i.e., the concentration of species on the catalyst outside surface is the same as that in the bulk gas phase.

For calculation of the vector of the reaction rates $\mathbf{r}$ in Eq. 5—5, the kinetic models of Graaf *et al.* [91] and Bercic & Levec [92] were applied to simulate the different reactions in the core and shell regions, respectively. In the model of Graaf *et al.* [91], methanol synthesis from CO, methanol synthesis from CO_2, and water gas shift reaction are considered:

Reaction 1: $CO + 2H_2 \leftrightarrow CH_3OH$,
Reaction 2: $CO_2 + 3H_2 \leftrightarrow CH_3OH + H_2O$,
Reaction 3: $CO + H_2O \leftrightarrow CO_2 + H_2$.

The kinetic expressions for reactions 1-3 are as follows [91]:

$$r_{CO} = \frac{k_1 K_5 (p_{CO} p_{H_2}^{3/2} - p_{CH_3OH} /(p_{H_2}^{1/2} K_{eq,p1}))}{(1 + K_5 p_{CO} + K_6 p_{CO_2}) \left[p_{H_2}^{1/2} + (K_7) p_{H_2O} \right]}, \qquad \text{Eq. 5—10}$$

$$r_{CO_2} = \frac{k_2 K_6 (p_{CO_2} p_{H_2}^{3/2} - p_{CH_3OH} p_{H_2O} /(p_{H_2}^{3/2} K_{eq,p2}))}{(1 + K_5 p_{CO} + K_6 p_{CO_2}) \left[p_{H_2}^{1/2} + (K_7) p_{H_2O} \right]}, \qquad \text{Eq. 5—11}$$

$$r_{RWGS} = \frac{k_3 K_6 p_{H_2} p_{CO_2} (1 - p_{H_2O} p_{CO} / p_{H_2} p_{CO_2} K_{eq,p3})}{(1 + K_5 p_{CO} + K_6 p_{CO_2}) \left[p_{H_2}^{1/2} + (K_7) p_{H_2O} \right]}. \qquad \text{Eq. 5—12}$$

For methanol dehydration:

Reaction 4: $2CH_3OH \leftrightarrow DME + H_2O$,

the following expression from Bercic & Levec [92] is used:

$$r_{DME} = \frac{k_4 K_8^2 (c_{CH_3OH}^2 - c_{H_2O} c_{DME} / K_{eq,p4})}{(1 + 2\sqrt{K_8 c_{CH_3OH}} + K_9 c_{H_2O})^4}. \qquad \text{Eq. 5—13}$$

The equilibrium constants of reactions 1-4 were calculated according to the literature [104]-[105] as follows:

$$\log_{10} K_{eq,p1} = \frac{5139}{T} - 12.621,$$ Eq. 5—14

$$\log_{10} K_{eq,p2} = \frac{3066}{T} - 10.592,$$ Eq. 5—15

$$\log_{10} K_{eq,p3} = \frac{-2073}{T} + 2.029,$$ Eq. 5—16

$$\log_{10} K_{eq,p4} = \frac{2167}{T} - 0.5194 \log_{10} T + 1.037 \times 10^{-3} T - 2.331 \times 10^{-7} T^2 - 1.2777$$ Eq. 5—17

According to the component material balances for the bulk gas phase in one discretized section, as shown in Figure 5-6 middle, the following equation applies:

$$N_{in,i} - N_{out,i} - N_{cat,i} = 0.$$ Eq. 5—18

$N_{cat,i}$: Flux of species *i* into or out of the core-shell catalyst particles, which are packed in this reactor section.

Finally, assembling all the above equations according to the component material balances, a nonlinear system with $(n_{shell} + n_{core} + 2) \times 6$ equations is obtained:

$$\mathbf{F(c)} = \mathbf{0}.$$ Eq. 5—19

Here, $n_{shell} + n_{core} + 2$ nodes exist in the discretized double layer shown in Figure 5-6 bottom, which consist of n_{shell} nodes in shell, n_{core} nodes in core, one node in the bulk gas phase, and one interface node between shell and core. This interface node can be used to simulate the effect of blocking or not perfect connection of pores in the intermediate layer between core and shell, which has been studied by the CPNM in [53]. Since there are 6 reactive species in the DME synthesis from syngas, 6 nonlinear equations are obtained for every node according to the component material balance.

5.4.4 Model implementation

The simulation model for the tube reactor described in the last two subsections was implemented in Matlab®. The overall algorithm for solution is illustrated in Figure 5-7. Firstly, the program reads in all the parameters, e.g., diffusion and kinetic parameters, which are summarized in Table 5-1. After that, the species concentration along the reactor is calculated section by section. In order to obtain the gas flow and composition out of one reactor section, the nonlinear equation system in Eq. 5—19 is solved with the function 'fsolve' from the Matlab® Optimization Toolbox®, which was also used to solve the CPNM as described in Section 3.3. Its default algorithm – "trust-region dogleg" was chosen. For a detailed description of the mathematics, see Powell [106]. As mentioned above, the obtained gas flow and composition out of this reactor section is used as the feed gas flow and composition of the next reactor section. Finally, the gas flow and composition in the outlet of the reactor are obtained which are then used to calculate conversion and selectivity. Moreover, the validity of the results obtained by the 'fsolve' routine was confirmed by starting from different initial values.

Table 5-1: Parameters used in the simulation model, adopted from [84].

Surface Diffusion Model						
Parameters	**CO_2**	**H_2**	**CH_3OH**	**H_2O**	**CO**	**DME**
$K_{523\,K}$ 1/Pa	3.5×10^{-7} [98]	1.5×10^{-8} [99]	1.3×10^{-6} [101]	8.4×10^{-6} [101]	4.3×10^{-7} [102]	$^{*}1.2\times10^{-6}$ [103]
E_a (K) kJ/mol	-28.7 [98]	-3.8 [99]	-19.65 [101]	-31.28 [101]	-15.9 [102]	*-33.3 [103]
q_{sat} mol/kg	2.14 [98]	9.5 [99]	2.6 [101]	2.68 [101]	1.1 [102]	*2.0 [103]
$D^{S}_{523\,K}$ m²/s	6.0×10^{-9} [96]	2.3×10^{-7} [100]	$^{+}6\times10^{-9}$ [97]	1.4×10^{-8} [97]	2.5×10^{-8} [96]	$^{+}6\times10^{-9}$ [97]
$E_a(D^S)$ kJ/mol	4.50 [96]	3.1 [100]	$^{+}$10 [97]	16.94 [97]	5.00 [96]	$^{+}$10 [97]

*: Adsorption parameters of C_2H_6 were used for DME.

$^{+}$: Diffusion parameters of CH_4 were used for methanol and DME.

Reactor Model					
T °C	250-280	p bar	40	L_{cat} m	0.095
$d_{p,\,core}$ µm	72	S_{shell} µm	3	$n_{reactor}$	1000
n_{core}	4	n_{shell}	10	$\bar{q}_{\mathrm{sat}}$ mol/kg	2.1
m_{cat} g	6	ρ_{core} kg/m³	4000	ρ_{shell} kg/m³	1500
d_p^{core} nm	25	$d_{p,meso}^{shell}$ nm	10	$d_{p,micro}^{shell}$ nm	0.56
ε_p^{core}	0.5	$\varepsilon_{meso}^{shell}$	0.001	$\varepsilon_{micro}^{shell}$	0.1
τ_p^{core}	2	τ_{meso}^{shell}	2	τ_{micro}^{shell}	2

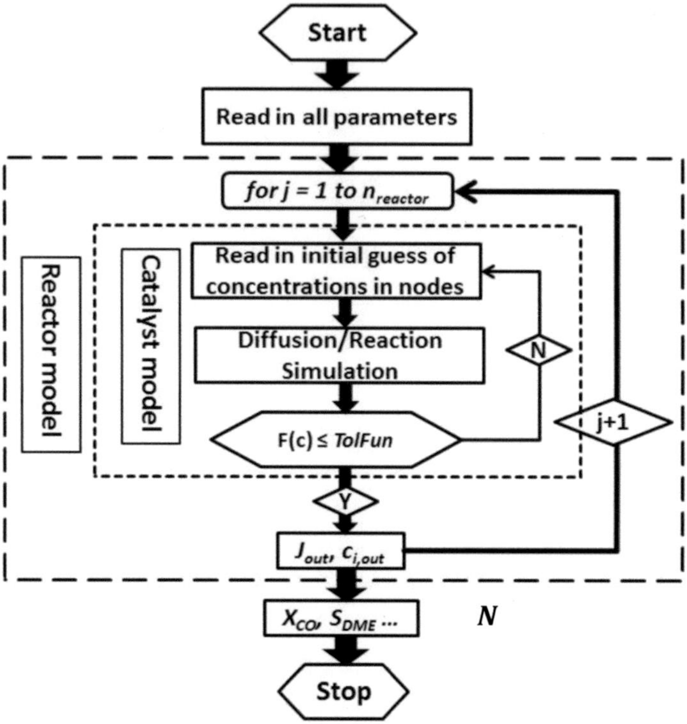

Figure 5-7: Program work flow for simulation of the tubular reactor packed with core-shell catalyst particles, adopted from [84].

5.5 Results and discussion

5.5.1 Model verification and determination of kinetic parameter estimates

In a previous study, Tauro [83] has used the same tubular reactor filled with a mixed bed of a commercial $Cu/ZnO/Al_2O_3$ catalyst and a self-synthesized

ZSM-5 catalyst for determination of kinetic parameter estimates in Eq. 5—10 - Eq. 5—13. The results are summarized in Table 5–2.

The parameters for the diffusion model and reactor model in Table 5-1 were used in the simulations. Since the same $Cu/ZnO/Al_2O_3$ catalyst was used as the core in the core-shell catalyst of the present study, and the ZSM-5 layer on the core catalyst was synthesized having the same Si/Al ratio as shown in [83], it was expected that these kinetic parameters could be used to simulate the behavior of the core-shell catalyst. However, the simulation results obtained with the kinetic parameters determined by Tauro [83] show the equilibrium CO conversion and DME selectivity at the reactor outlet, much higher than those determined experimentally (see Figure 5-5). Lower CO conversion of the core-shell catalyst could have been caused by diffusion limitation in shell or by a lower activity of the core catalyst. However, the increasing selectivity to DME and decreasing selectivity to methanol with increasing temperatures implies that the methanol produced in the core was not converted fast enough to DME, i.e., the shell catalyst was not sufficiently active.

In order to investigate this behavior, keeping the activation energies at the same values as identified in [83], the reaction rates k_1 - k_4 were adjusted to match the experimental data, *via* multiplying the rate constants by the correction factors f_{core} and f_{shell}, while all the other parameters were kept as in [83]. Thus, the modified reaction rates can be calculated as:

$$\begin{cases} k_i' = k_i \cdot f_{core} & i = 1,2,3 \\ k_i' = k_i \cdot f_{shell} & i = 4 \end{cases} . \qquad \text{Eq. 5—20}$$

Table 5-2: Kinetic parameters determined via fitting of the experimental data, adopted from [84]. k_i or $K_i = k_0 \cdot \exp(-E_a / RT)$

	Tauro [11]		**This study**		
	k_0	E_a [J/mol]	k'_0	E_a [J/mol]	
k_1	$4.79 \cdot 10^{6}$	102160	$\mathbf{9.58 \cdot 10^{5}}$	102160	$[mol \cdot kg_{cat}^{-1} \cdot s^{-1} \cdot bar^{-1}]$
k_2	$1.39 \cdot 10^{6}$	95209.8	$\mathbf{2.78 \cdot 10^{5}}$	95209.8	$[mol \cdot kg_{cat}^{-1} \cdot s^{-1} \cdot bar^{-1}]$
k_3	$1.06 \cdot 10^{12}$	135156	$\mathbf{2.12 \cdot 10^{11}}$	135156	$[mol \cdot kg_{cat}^{-1} \cdot s^{-1} \cdot bar^{-0.5}]$
k_4	$6.21 \cdot 10^{11}$	94197.8	$\mathbf{2.48 \cdot 10^{8}}$	94197.8	$[mol \cdot kg_{cat}^{-1} \cdot s^{-1}]$
K_5	$7.99 \cdot 10^{-7}$	-55300	$7.99 \cdot 10^{-7}$	-55300	$[bar^{-1}]$
K_6	$1.02 \cdot 10^{-7}$	-70000	$1.02 \cdot 10^{-7}$	-70000	$[bar^{-1}]$
K_7	$4.13 \cdot 10^{-11}$	-104500	$4.13 \cdot 10^{-11}$	-104500	$[bar^{-0.5}]$
K_8	$7.90 \cdot 10^{-7}$	-75000	$7.90 \cdot 10^{-7}$	-75000	$[m^3 \cdot mol^{-1}]$
K_9	$8.40 \cdot 10^{-5}$	-41000	$8.40 \cdot 10^{-5}$	-41000	$[m^3 \cdot mol^{-1}]$

In the fitting simulations, at first, the calculations with the combined factors f_{core} (0.1, 0.25, 0.4, 0.55, 0.7, 0.85, 1) and f_{shell} (10^{-5}, 5×10^{-5}, 10^{-4}, 5×10^{-4}, 10^{-3}, 5×10^{-3}, 10^{-2}, 5×10^{-2}, 0.1, 0.5, 1) showed that the sum of the relative deviations from the experimental data of CO conversions, selectivities to DME and methanol was the smallest, when $f_{core} = 0.55$ and $f_{shell} = 5\times10^{-4}$. After that, the intervals of f_{core} [0.4, 0.7] and of f_{shell} [10^{-4}, 10^{-3}] were equally divided into 6 and 10 intervals in the calculations. Finally, it was found that the best fitting f_{core} and f_{shell} were 0.5±0.05 and $(4\pm1)\times10^{-4}$, respectively. The determined kinetic parameters for the core-shell catalyst are summarized in Table 5–2. In Figure 5-8, the best fitting conversion of CO, and selectivities to DME and methanol are shown.

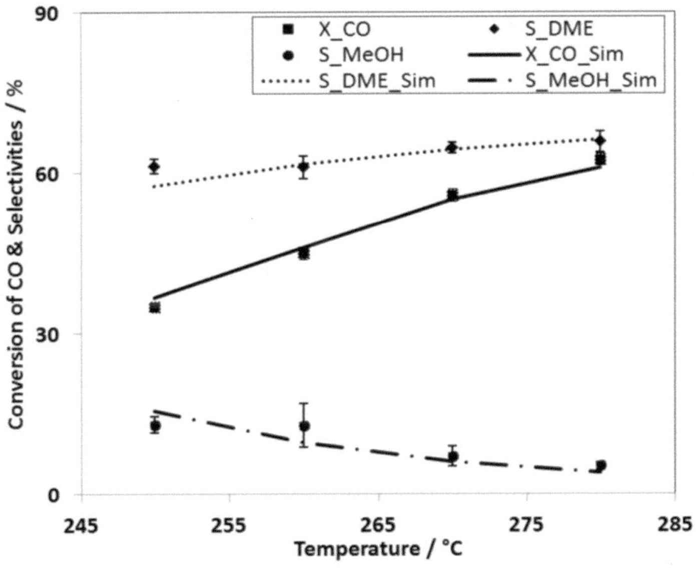

Figure 5-8: Simulated CO conversions and product selectivities in DME direct synthesis on a core-shell catalyst in a tubular reactor with the best fitting correction factors $f_{core} = 0.5$ and $f_{shell} = 4\times10^{-4}$, at 250-280 °C. Operating conditions: 40 bar, GHSV: 900 $ml_{STP}/g_{cat}/h$, N_2:H_2:CO:CO_2 = 10:57.6:28.8:3.6, adopted from [84].

The small values of the correction factors for the shell and core f_{shell} = $(4\pm1)\times10^{-4}$ (close to zero), f_{core} = 0.5±0.05 imply that the H-ZSM-5 shell catalyst did reach a much lower activity than the H-ZSM-5 catalyst prepared by Tauro [83], while the core catalyst has a slight deactivation. The deactivation of the zeolite shell may be caused by the organics existing in the core catalyst. During the calcination, the organics diffuse out from the core and block the zeolite pores in the shell. During the catalyst preparation, the zeolite crystals may grow in the pores of the core, which leads to the deactivation of the core. In Figure 5-9, the molar fraction profiles of all species simulated with the original kinetic parameters from [83] (dashed curves) show that the direct DME synthesis reaches equilibrium at the catalyst bed length of 0.025 m, whereas the solid lines show that equilibrium is not achieved in case of simulation with the corrected kinetic parameters determined here.

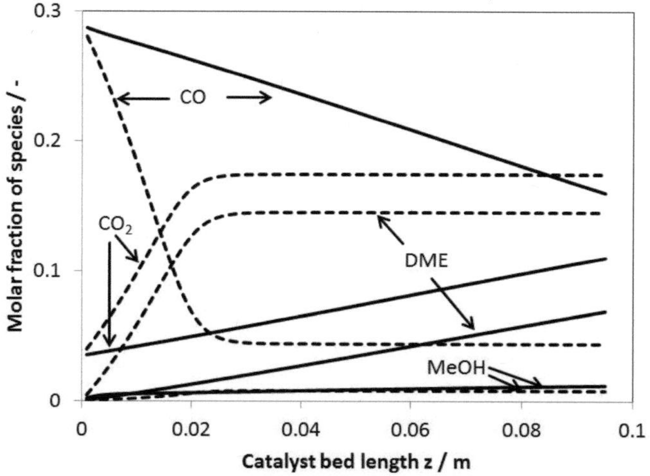

Figure 5-9: Simulated molar fraction profiles of species along the catalyst bed in the tubular reactor at 270 °C. Solid curves: with correction factors f_{core} and f_{shell} of 0.5 and 4×10^{-4}; Dashed curves: without correction (Uncorrected), adopted from [84].

5.5.2 Influence of shell thickness and activity

The combined effect of the shell thickness and activity was investigated *via* varying the shell thickness from 1 to 8.5 µm and the activity factor f_{shell} from $4 \cdot 10^{-4}$ to 0.4.

Figure 5-10 shows that the conversion of CO and the selectivity to DME both increase significantly with increasing shell thickness and activity. This agrees qualitatively with experimental results reported in [90], showing that the acidity of the shell catalyst and the thickness of the shell layer both have significant effect on the catalytic performance of a bifunctional $Cu/ZnO/Al_2O_3@SiO_2/Al_2O_3$ core-shell catalyst.

Moreover, Figure 5-10 shows that when the correction factor f_{shell} for k_4 is above 0.04, i.e., the shell catalyst has more than 4 % activity of the H-ZSM-5 catalyst prepared by Tauro [83], the maximum CO conversion and DME selectivity are achieved by the core-shell catalyst even with a 1 µm thick shell. Thus, it suggests to achieve a sufficiently active and thin shell catalyst by a suitable preparation method, so that it can completely convert the methanol produced by core, but has not serious diffusion limitation on the reactants of the methanol synthesis, i.e., CO and H_2.

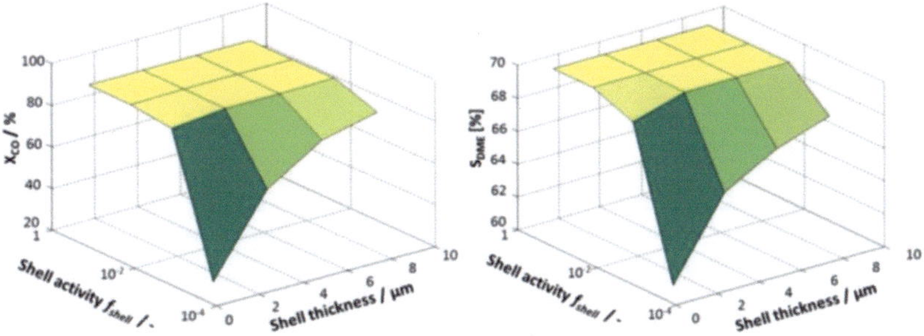

Figure 5-10: Simulated CO conversion (left) and selectivity to DME (right) as a function of the shell thickness and activity (i.e., fshell) at T = 270 °C, p = 40 bar, GHSV= 900 mlSTP/gcat/h and defect porosity = 10^{-3}, adopted from [84].

6 Crystallite-pore network model for core-shell catalysts

Since the CPNM can investigate the effect of the polycrystalline and anisotropic structure of the zeolites (e.g., the concentration of defect pores, the crystallite orientation), a simulation model based on the CPNM presented in Chapter 3, a more detailed simulation model than the heterogeneous model in Chapter 5, has been developed to simulate diffusion and reaction in a single core-shell catalyst particle placed in a perfectly mixed reactor (or reactor section), e.g., a Berty-type reactor (see Figure 6-1 left). In this chapter, the development of the simulation model based on the CPNM and results of the core-shell catalyst optimization for one-stage DME synthesis are shown.

6.1 Development of a model based on the CPNM

6.1.1 Core-shell catalyst in a perfectly mixed reactor for catalytic test

Besides the tubular reactor, a perfectly mixed Berty-type reactor available at IMVT will be used in an ongoing PhD thesis to test the catalytic performance of the prepared core-shell catalysts. A schematic of the Berty-type reactor placed with the core-shell catalysts is shown in Figure 6-1, a labeled photograph can be found in Appendix c). In contrast to the integral tubular reactor, the Berty-type reactor is free of concentration gradients in gas phase due to high recycle gas velocity inside created by stirring, therefore performs as an ideal continuous stirred-tank reactor (CSTR). Thus, it can investigate the catalyst performance excluding the influence of the concentration gradient outside the catalyst particle.

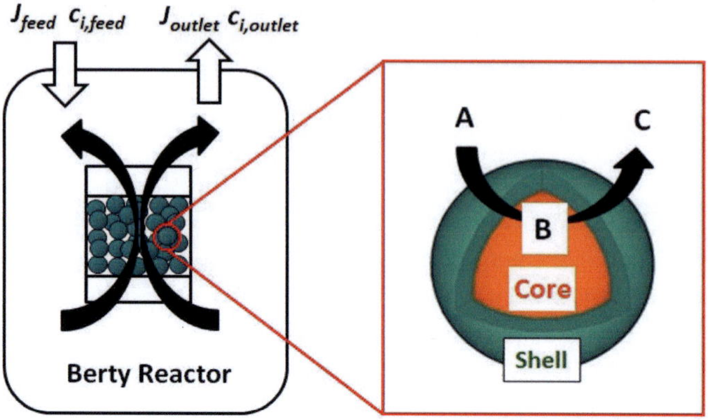

Figure 6-1: Schematic of a Berty-type reactor (left) placed with the core-shell catalyst particles (right).

6.1.2 Transformation of core-shell to double layer structure

A simulation model based on the CPNM was developed to simulate one-stage DME synthesis in the core-shell catalyst in the Berty-type reactor. Since the transport resistance in the microporous shell layer is dominant due to low diffusivities of gases in this layer, it is allowed to assume absence of concentration gradients in the core catalyst. This assumption has been evaluated *via* the simulation of the heterogeneous model in Section 5.4.3. Thus, the core-shell structure is transformed to a double-layer structure in the model based on the CPNM as in the heterogeneous model in Chapter 5 for simplicity (see Figure 6-2 left). The CPNM was used to simulate the polycrystalline zeolite shell layer (see Figure 6-2, right). In the simulation model, the exterior layer of the double-layer structure is as thick as the shell layer of the core-shell catalysts, while the thickness of the interior layer is one-sixth of the core diameter to keep the same volume ratio of core and shell catalysts as for the spherical geometry. In the following sub-sections, the mathematical model and its implementation in Matlab® is introduced.

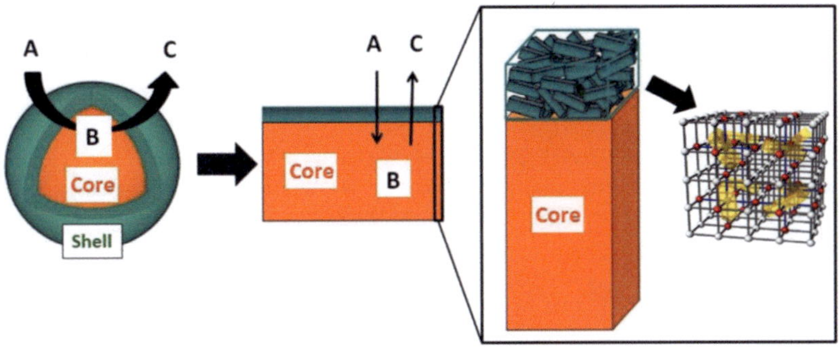

Figure 6-2: Left: Transformation of core-shell into double-layer structure;
Right: crystallite-pore network representing the polycrystalline zeolite shell layer.

6.1.3 Simulation model based on the CPNM for core-shell catalyst

The structural parameters used in the 3-D simulation model based on the CPNM in the base case are listed in Table 6-1. In this model, diffusion in the defects (intercrystalline pores) and intracrystalline pores of the shell was simulated by the KDM und SDM, respectively. Moreover, KDM was applied to model the diffusion in the mesopores of the core catalyst. The concentration gradients in the core catalysts were simulated by the discretization method, which has been used in the heterogeneous model in Chapter 5. The adsorption parameters of gases CO_2, water, methanol and DME in ZSM-5 determined with the LCM adsorption measurement device were used in SDM to simulate the diffusion in the zeolite micropores, while other adsorption and diffusion parameters were obtained from literature (see Table 5-1). Moreover, it was assumed that the reaction in the shell (i.e., methanol dehydration) takes place only in the ZSM-5 crystallites, i.e., in the nodes representing the crystallites. The kinetic models of Graaf *et al.* [91] and Bercic & Levec [92] used in the heterogeneous model in Chapter 5 were also applied in this model to simulate the reactions in the core and shell layers, respectively. Considering that the core and shell catalysts were not deactivated significantly during the catalyst preparation, the kinetic parameters determined by Tauro [11] (see Table 5–2) were used in these kinetic models.

Table 6-1: Structural parameters used in simulation model in base case

Parameters		**Determined or**	**Estimated**	**Sources**
Crystallite length / μm		2	---	[84]
Shell thickness / μm		6	---	[84]
Area of catalyst surface / μm×μm		---	4×4	*
Diameter of core particle / μm		72	---	[84]
Discretization of the core layer		---	8 nodes	*
Pore in shell / nm	defects	---	5	*
	zeolite pore	0.55	---	[107]
Pore in core / nm		25	---	**
Connectivity of non-zeolite pore		---	4.2	*
Crystallite orientation	ψ	---	45°	*
	χ	---	Rand(1)×360°	*
Density of shell / kg m^{-3}		1835	---	**
Density of core / kg m^{-3}		3970	---	**
Porosity of shell	defects	---	10^{-7}	*
	zeolite pore	---	0.1	[108]
Porosity of core		---	0.5	*

* Investigation of sensitivity; ** Measured at CRT. Rand(1): Matlab function, which returns a pseudorandom value drawn from the standard uniform distribution on the open interval (0, 1).

The core-shell catalyst was thought to be placed in the Berty-type reactor (or a perfectly mixed reactor section as shown in Figure 5-6 middle), which 3 was free of concentration gradients in gas phase inside (i.e., outside the catalyst particles). The operating conditions of this reactor were given by the contact time W/F (catalyst mass/molar feed flow) = 8×10^{-6} kg_{cat} h mol^{-1}, the temperature T = 270 °C, and the pressure p = 40 bar. The feed gas has the same composition as that used in the test by the tubular reactor (H_2: CO: CO_2: N_2 = 57.6 %: 28.8 %: 3.6 %: 10 %) in Chapter 5.

6.2 Results and discussion

6.2.1 Simulated concentration distribution in the CPNM

Figure 6-3 shows the simulated concentration distribution of the product DME in the CPNM which represents the shell layer. The top and bottom plane of this network was assumed to be in contact with the core catalyst and the bulk gas phase, respectively. It can be seen that the concentration of DME in the top plane (about 1 %) is higher than that in the bottom plane contacting with the bulk gas phase (about 0.5 %), while the concentration in the crystallite node is slightly higher than that in the interface nodes close to this crystallite node. It agrees with the fact that the methanol from the core is converted to DME in the zeolite crystallites. Afterwards, the DME diffuses out of the zeolite crystallites into the gas phase *via* the interface nodes and defect pores due to the concentration gradients.

6.2.2 Effect of defects

The porosity of the defect pores (i.e., non-zeolite pores) in the H-ZSM-5 shell layer, which represents the area fraction of the defect pores in the shell surface, varied from 10^{-7} to 0.1. The results shown in Figure 6-4 indicate that the porosity of the defect pores has an obvious influence on the catalytic performance of the core-shell catalyst. While the porosity of defects decreases from 0.1 to 10^{-5}, the conversion of CO decreases and the selectivity to DME increases significantly. This can be explained that when the concentration of defects in shell decreases, more and more gases have to diffuse through the shell *via* the zeolite micropores. This leads on one hand to lower conversion of CO due to low diffusivity of synthesis gas in the zeolite micropores, and on the other hand to higher selectivity towards DME because of longer contact time of methanol produced in the core on the way back to the gas phase. However, the conversion of CO and the selectivity to DME level off, when the porosity of defects is below 10^{-5}. Under this condition, almost all the gases diffuse through the shell *via* the zeolite micropores. Thus, the shell layer is practically defined defect-free when the porosity (i.e., area fraction) of defects is below 10^{-5}, i.e., 0.001 %.

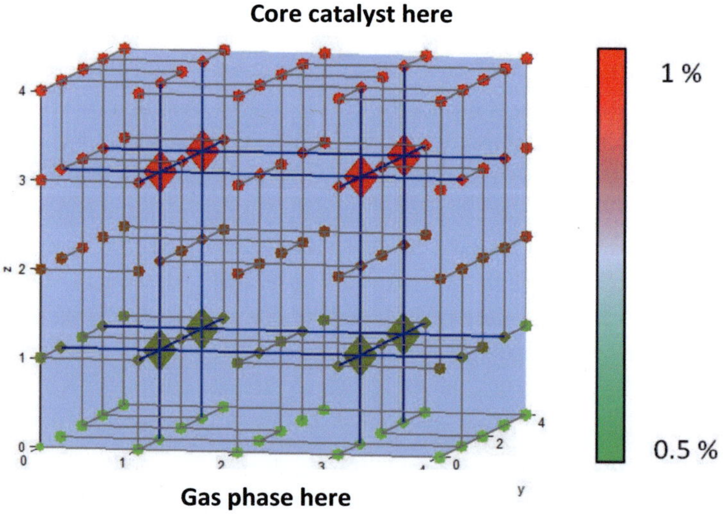

Figure 6-3: Simulated concentration distribution of DME in the CPNM in base case (Conditions as of Table 6-1).

Finally, the simulations give the suggestion that the concentration of defects in the shell should be controlled by appropriate actions during membrane preparation or a post synthetic treatment to reduce it afterwards for an optimal catalyst performance considering the conversion of CO and selectivity to DME simultaneously.

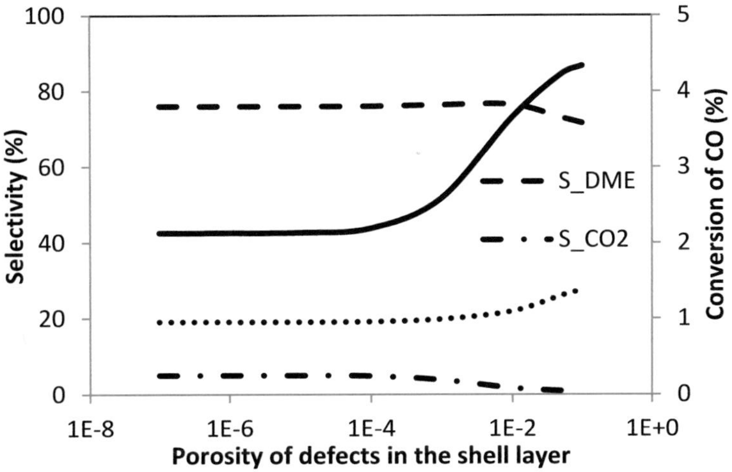

Figure 6-4: Dependence of CO conversion (X_CO) and product selectivities on porosity of defects (Conditions as of Table 6-1).

6.2.3 Effect of crystallite orientation

In order to investigate the effect of crystallite orientation on the catalytic performance of the core-shell catalyst, the angle ψ between the *b*-direction of the crystallite and the surface of the core catalyst (see Figure 3-2) varied from 10° to 90°. A larger angle ψ means that the shell layer is more *b*-oriented (the zeolite pores in *b*-direction run parallel to the shell thickness), i.e., the shell layer has higher diffusivity. The simulations show that the crystallite orientation has a significant effect on the catalytic performance of the core-shell catalyst, only when the shell is defect-free, i.e., the porosity of defects is below 0.001 %. Figure 6-5 shows the simulation results indicating that the orientation of the ZSM-5 crystallites in the defect-free shell can influence the conversion of CO and the selectivity to DME significantly. This is because an increasing angle ψ reduces the transport resistance of the ZSM-5 layer. Hence, the conversion of CO increases with increasing angle ψ. However, the selectivity towards DME decreases with increasing angle ψ, as the contact time of methanol inside the shell gets reduced as well.

Same as the concentration of defects in the shell, the optimal crystallite orientation in the shell can be identified considering the conversion of CO and the selectivity towards DME simultaneously.

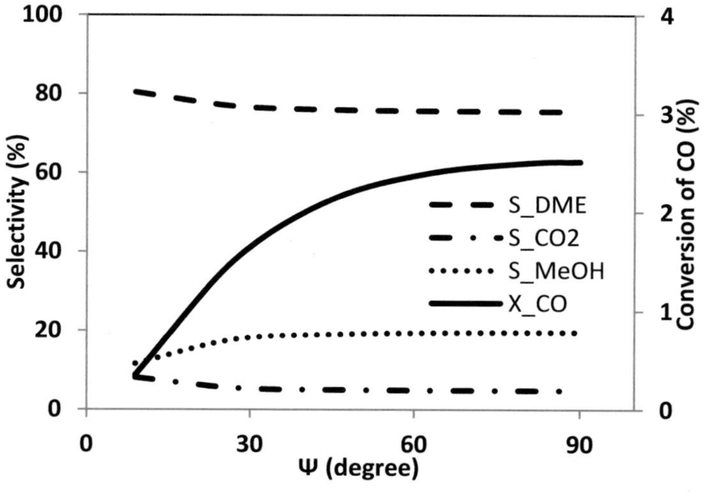

Figure 6-5: Dependence of CO conversion (X_CO) and product selectivities on crystallite orientation of the defect-free shell (Conditions as of Table 6-1).

6.2.4 Effect of shell thickness

The effect of the shell thickness was investigated by varying from 0.5 to 2.5 µm, while keeping the crystallite size of 0.5×0.5×0.5 µm^3. The simulation results in Figure 6-6 indicate that the catalytic performance of the core-shell catalyst is affected by the shell thickness significantly. With increasing the shell thickness, the conversion of CO decreases significantly due to the increasing transport resistance, while the selectivity towards DME has a significant increase when the thickness is below 1.5 µm owing to the increasing contact time of methanol inside the shell. However, the selectivity towards DME increases with increasing the shell thickness not so significantly, when the shell is thicker than 1.5 µm. This can be explained that the methanol produced in the core can be converted to DME by the 1.5 µm thick shell almost completely. In this case, increasing the shell thickniss leads only to a lower CO conversion. Therefore, the core-shell catalyst is suggested to have a very reactive but thin shell, which can convert the methanol from core completely to DME, but has small transport resistance.

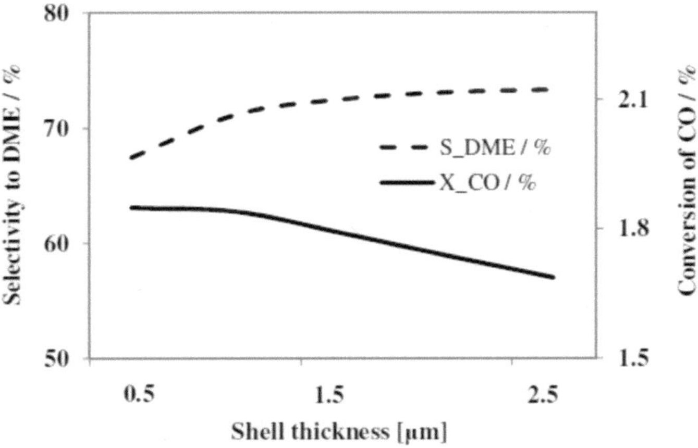

Figure 6-6: Dependence of CO conversion (X_CO) and selectivity towards DME on the thickness of the defect-free shell (crystallite size of 0.5×0.5×0.5 µm^3, other conditions as of Table 6-1).

6.3 Conclusions

The simulation model based on the CPNM has been developed to predict the catalytic performance of the core-shell catalyst for direct DME synthesis from synthesis gas on the particle level. In this model, the 3-D crystallite-pore network model was applied to simulate the polycrystalline microporous structure of the ZSM-5 zeolite layer. Effect of the polycrystalline and anisotropic structure of the zeolite shell layer (e.g., concentration of defect pores, crystallite orientation), as well as the shell thickness, has been studied. Moreover, this simulation model can help us to understand the complex transport and reaction processes occurring in the core-shell catalyst. Some guidances on the catalyst preparation and optimization were also given from the simulation results, e.g., controlling the concentration of defect pores in the shell, optimizing the orientation of zeolite crystallites considering the CO conversion and the selectivity towards DME simultaneously, and synthesizing a very reactive but thin shell.

7 Crystallite-pore network model for xylene isomerization in ZSM-5 membrane

With increasing demand for *p*-xylene as a starting material for the manufacture of synthetic fibers (polyester), efforts are being directed to selectively produce *p*-xylene from low valued *m*- and *o*-xylenes. Unfortunately, the amount of *p*-xylene theoretically obtainable from its isomers by isomerization is limited by thermodynamics. In order to overcome the limitation of thermodynamics, a xylene transformation process using ZSM-5 membranes was introduced by Mobil in the early 1970s, which represented an obvious technological improvement. Since ZSM-5 has intracrystalline pores with a diameter of 0.56 nm (kinetic diameter: *p*-xylene 0.58 nm, *m*- and *o*-xylene 0.68 nm) and active sites for xylene isomerization, catalytic ZSM-5 membranes are expected to provide good activity for xylene isomerization combined with selective separation of *p*-xylene from its bulkier isomers. Moreover, a *b*-oriented ZSM-5 membrane prepared by Lai, et al. (2003) [16] showed better permeation and separation performance for xylene isomers than randomly oriented membranes. This implies that the orientation of ZSM-5 crystallites in the membrane has a significant effect on its catalytic performance in xylene isomerization. Thus, the xylene isomerization in a catalytic ZSM-5 membrane reactor is another interesting case for the CPNM presented in Chapter 3.

In this chapter, as the second test example for the CPNM, xylene isomerization in a catalytic ZSM-5 membrane prepared by Haag and co-workers (2006) [13] was simulated with a simulation model based on the CPNM. At first, diffusion and adsorption parameters of xylene isomers in the ZSM-5 membrane were determined from single gas permeation data reported in that paper. Afterwards, using the kinetic expression for xylene isomerization and the experimental data of the catalyst tests from [13], kinetic parameters were obtained with the simulation model based on the CPNM by regression. Finally, effects of some selected structural parameters of the membrane were investigated.

7.1 Simulation model development

Haag et al. [13] used a similar configuration like the WK cell introduced in Section 4.2 as a catalytic membrane reactor to test the performance of ZSM-5 membranes in xylene isomerization. In the single gas permeation and xylene isomerization experiments, only single-isomer feeds were studied. Details on the experiments, the experimental setup and the operating conditions can be found in Ref. [13].

In the simulation model for xylene isomerization in the ZSM-5 membrane placed in the WK cell, the crystallite-pore network model was applied to simulate diffusion and reaction in the polycrystalline ZSM-5 membrane, as shown in Figure 7-1. In this model, xylene isomerization was assumed to take place only inside the crystallites. Moreover, the single-file diffusion model (SFDM) was used to simulate the diffusion of the xylene isomers in the intracrystalline micropores, which have a molecular size larger than the micropore size, while the Knudsen diffusion model (KDM) was again used to simulate the diffusion through the defects in the membrane.

The zeolite membrane was synthesized on a stainless steel support layer. The thickness was 2 mm, and the fine-porous skin layer had an average pore diameter of about 250 nm [13]. Thus, due to much larger pores in the stainless steel support layer, the transport resistance in the support layer was assumed to be neglected in the simulation model.

In the simulation model, the flows of species i through the transport boundary layers from permeate and retentate chambers into the outside nodes of the CPNM were calculated by:

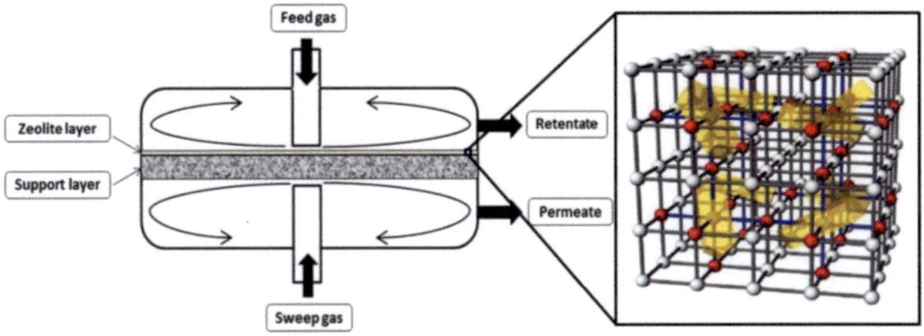

Figure 7-1: Wicke-Kallenbach cell with a supported zeolite membrane, which is simulated with the CPNM. The support layer is a stainless steel disk, more information can be found in [13].

Permeate:

$$\begin{cases} N'^{per}_{inter,i} = \varsigma_{inter} \cdot A_{inter} \cdot \beta_i^{per}(c_i^{per} - c_{outside,i}^{per}) \\ N'^{per}_{intra,i} = \varsigma_{intra} \cdot A_{intra} \cdot \beta_i^{per}(c_i^{per} - c_{outside,i}^{per}) \end{cases},$$ Eq. 7—1, Eq. 7—2

Retentate:

$$\begin{cases} N'^{ret}_{inter,i} = \varsigma_{inter} \cdot A_{inter} \cdot \beta_i^{ret}(c_i^{ret} - c_{outside,i}^{ret}) \\ N'^{ret}_{intra,i} = \varsigma_{intra} \cdot A_{intra} \cdot \beta_i^{ret}(c_i^{ret} - c_{outside,i}^{ret}) \end{cases}.$$ Eq. 7—3, Eq. 7—4

where β_i is the mass transfer coefficient of species i. Since only single-isomer feeds were used in the experiments, the transport resistance in the boundary layers is generally neglectable compared to the dominant transport resistance in the zeolite membrane. In order to verify this, the mass transfer coefficients of 0.01, 0.1 and 1 m s^{-1} were used in the simulations and found no significant effect of inaccurate mass transfer coefficients on the simulation results, i.e. the mass transfer coefficients of xylene isomers are insensitive parameters. Thus, a uniform value (0.01 m s^{-1}) was given to the mass transfer coefficients of xylene isomers in the present model.

Local blocking, e.g., due to imperfect match of the pore systems of the zeolite and support layer, was considered by introducing the proportion of blocked pores b ($0 \le b \le 1$) in the following equations:

Permeate:

$$\begin{cases} N'^{per}_{inter,i} = (1-b^{per}_{inter}) \cdot \varsigma_{inter} \cdot A_{inter} \cdot \beta^{per}_i (c^{per}_i - c^{per}_{outside,i}) & \text{Eq. 7—5} \\ N'^{per}_{intra,i} = (1-b^{per}_{intra}) \cdot \varsigma_{intra} \cdot A_{intra} \cdot \beta^{per}_i (c^{per}_i - c^{per}_{outside,i}) & \text{Eq. 7—6} \end{cases},$$

Retentate:

$$\begin{cases} N'^{ret}_{inter,i} = (1-b^{ret}_{inter}) \cdot \varsigma_{inter} \cdot A_{inter} \cdot \beta^{ret}_i (c^{ret}_i - c^{ret}_{outside,i}) & \text{Eq. 7—7} \\ N'^{ret}_{intra,i} = (1-b^{ret}_{intra}) \cdot \varsigma_{intra} \cdot A_{intra} \cdot \beta^{ret}_i (c^{ret}_i - c^{ret}_{outside,i}) & \text{Eq. 7—8} \end{cases}.$$

According to the component material balances in the retentate and permeate spaces, one obtains:

Permeate: $N'_{sweep} y_i^{sweep} - N'_{per} y_i^{per} - \Sigma N'^{per}_i = 0$, Eq. 7—9

Retentate: $N'_{feed} y_i^{feed} - N'_{ret} y_i^{ret} - \Sigma N'^{ret}_i = 0$. Eq. 7—10

Finally, assembling all the equations for nodes in the CPNM network and spaces according to the component material balances, a large system of nonlinear equations was obtained:

$$\mathbf{F(c)} = \mathbf{0},$$ Eq. 7—11

which was implemented in Matlab® and solved with the function 'fsolve'. More details about how to implement and solve the simulation model presented here can be found in Chapter 3.

7.2 Simulation results and discussion

7.2.1 Estimation of diffusion and adsorption parameters

For estimation of the diffusion and adsorption parameters of the xylene isomers, the geometric and operating parameters given in the paper [13] by Haag et al. were used. The ZSM-5 membrane used by Haag et al. [13] had a surface area of 2.54 cm^2 and a zeolite layer thickness of 6 μm. The SEM image shows that the ZSM-5 crystallites in the membrane had a size of ≈ 1×2×3 μm^3. Thus, the unit pore length in the CPNM network as shown in Figure 3-2 was set to 1 μm. At a pressure of 1 bar, single gas permeation of xylene isomers was studied by Haag et al. [13] at 100 °C and 200 °C, respectively. The feed was supplied in liquid form at a flow rate of 0.0574 ml min^{-1} (20 °C) and was evaporated at 170 °C. A nitrogen carrier gas flow of 50 ml min^{-1} (STP) was used to transport the evaporated feed to the retentate compartment. A nitrogen flow of 250 ml min^{-1} (STP) served as permeate sweep gas. Besides these data, other parameters adopted from the paper by Haag et al. [13] and assumed/estimated parameters used in the simulation model are summarized in Table 7-1.

Table 7-1: Pore network properties used in the standard simulation, adopted from [53].

Parameters	Values	Sources
Number of nodes	175 (5×5×7)	Zeolite thickness [13]
Number of nodes representing crystallites	12 (2×2×3)	Zeolite thickness [13]
Unit pore length	1 μm	Crystallite size [13]
Mean size of intercrystalline mesopores $d_{p,inter}$	12 nm	Assumed
Size of intracrystalline micropores $d_{p,intra}$	0.55 nm	[107]
Porosity of intercrystalline mesopores ξ_{inter}	9×10^{-3}	Estimated
Connection possibility of intercrystalline pores P	100 %	Assumed
Porosity of intracrystalline micropores ξ_{intra}	0.2	[108]
mass transfer coefficient β	5	Assumed
Fraction of blocked intercrystalline pores b_{inter}^{per}	95 %	Estimated
Fraction of blocked intracrystalline pores b_{intra}^{per}	50 %	Estimated
Density of ZSM-5 membrane ρ	1990 kg m^{-3}	[13]
Angle ψ	45°	Randomly intergrown crystallites [13]
Angle χ	Rand(1)* × 360°	

* Rand(1): Matlab® function, which returns a pseudorandom value drawn from the standard uniform distribution on the open interval (0, 1). Estimated: chosen based on plausibility of parameter values and simulation results for single-gas permeation.

Table 7-2: Adsorption equilibrium constants and saturation loadings of xylene isomers at 100 °C, adopted from [53]

Adsorption parameters	*p*-xylene	*m*-xylene	*o*-xylene
$K_{i,373K}$ Pa^{-1}	$(1.00 \pm 0.10) \times 10^{-3}$	2.25×10^{-4} (assumed)	$(2.25 \pm 0.40) \times 10^{-4}$
q_{sat} mol/kg	0.693	0.693 (assumed)	0.693

Single-component adsorption isotherms for *p*- and *o*-xylene on a ZSM-5 membrane were measured by Xomeritakis et al. (2000) [109] at 100 °C. *Via* fitting their experimental data with Langmuir equations, the adsorption parameters of *p*- and *o*-xylene in the ZSM-5 membrane at 100 °C were determined. The obtained values (i.e., the adsorption equilibrium constants and saturation loadings) are shown in Table 7-2 (for the fitting results refer to Figures A7 and A8 in the Appendix d)). Unfortunately, adsorption isotherms for *m*-xylene were not measured by Xomeritakis et al. (2000) [109]. Therefore, the adsorption equilibrium constant of *o*-xylene at 100 °C was used as a first approximation also for *m*-xylene based on the similarity of their physical properties.

Based on the equation of SDM for a unary system Eq. 4—7 in Section 4.2.4 and using the parameters from Table 7-1 and adsorption parameters from Table 7-2, the corrected surface diffusion coefficients of the xylene isomers D_i^S at 100 and 200 °C were determined *via* fitting the predicted single gas permeances to the experimental data of Haag et al. [13]. The obtained parameter estimates are summarized in Table 7-3 and Table 7-4, whereas the predicted and experimental single gas permeances are shown in Figure A9 in Appendix d). For comparison with literature results, Table 7-3 also provides corrected surface diffusion coefficients (i.e., based on the chemical potential gradient) derived from zero-length-column (ZLC) studies by Brandani et al. [25] as well as uncorrected diffusion coefficients (i.e, based on the loading gradient) from uptake measurements by Zheng et al. [26].

It turns out that the corrected surface diffusion coefficients of the xylene isomers obtained here are rather close to each other whereas in literature in some cases up to 3 orders higher diffusion coefficients for *p*-xylene have been found compared to *m*-xylene and *o*-xylene. However, an interpretation of the differences has to consider the deviant definition of the two types of diffusion coefficients. The corrected surface diffusion coefficients D_i^S defined by Eq. 4—7 are based on the chemical potential gradient as the driving force for transport, whereas the uncorrected diffusion coefficients D_i are based on the loading gradient and are defined as follows:

$$J_i^S = -\rho D_i \nabla q_i = -\rho q_{sat} D_i \nabla \theta_i \,.$$

Eq. 7—12

Equating 7—13 and Eq. 4—7, and rearranging leads to:

$$D_i = D_i^S \Gamma = \frac{D_i^S}{1-\bar{\theta}_i} = D_i^S \cdot \frac{1}{1-{K_i\bar{p}_i}/{1+K_i\bar{p}_i}} = D_i^S \cdot (1+K_i\bar{p}_i),$$

Eq. 7—14

where Γ is the thermodynamic factor. A pronounced difference between the adsorption constants of *p*-xylene and the other isomers, as given in Table 7-2, can therefore cause a marked difference of the uncorrected diffusion coefficients of the xylene isomers even if the corrected diffusion coefficients are close to each other. The ratio of the corrected surface diffusion coefficients $D_p^s : D_m^s : D_o^s$ estimated from the data at 200 °C turned out as 6.59:2.85:1. This is not far from the ratio based on the results of zero-length-column studies by Brandani et al. [110], who found $D_p^s : D_m^s : D_o^s$ = 10:1.86:1 for 200°C.

On the other hand, comparing the absolute values shows that the corrected surface diffusion coefficients from the analysis here are orders of magnitude larger than those from literature. This may be caused by deviations of the assumed structural parameters from the real values, by inaccuracies of the adsorption constants of the xylene isomers, which have been adopted in the estimation of the corrected surface diffusion coefficients (in particular for *m*-xylene), and by a different size of the crystallites. Moreover, the present model ignores gradients of the surface coverage inside the crystallites, which may also contribute to errors in estimating the corrected surface diffusion coefficients.

Table 7-3: Corrected surface diffusion coefficients D_i^S of the xylene isomers at 100 °C and 200 °C obtained *via* fitting single gas permeance data from Haag et al. [13]; comparison to literature values, i.e., corrected surface diffusion coefficients at 200°C [110] and uncorrected diffusion coefficients at 100°C [111] , adopted from [53].

Diffusivities (m^2/s)	**T (°C)**	***p*-xylene**	***m*-xylene**	***o*-xylene**	$D_p^s : D_m^s : D_o^s$ / $D_p : D_m : D_o$
D_i^S	100	$(4.69 \pm 0.47) \times 10^{-10}$	$(2.20 \pm 0.33) \times 10^{-10}$	$(1.99 \pm 0.30) \times 10^{-10}$	2.36:1.10:1
(this work)	200	3.26×10^{-9}	1.41×10^{-9}	4.94×10^{-10}	6.59:2.85:1
D_i **[111]**	100	4.7×10^{-17}	3.8×10^{-19}	6.3×10^{-18}	7.46:0.06:1
D_i^S **[110]**	200	3.0×10^{-12}	5.6×10^{-13}	3.0×10^{-13}	10:1.86:1

The activation energies of D_i^S and K_i were determined *via* fitting the single gas permeance data of the xylene isomers at 200 °C by Haag et al. [13]. The results are shown in Table 7-4. The activation energies of D_i^S for *p*-, *m*-, and *o*-xylene are 28.8±1.6, 27.6±2.8 and 13.5±5.5 kJ/mol, respectively. They agree reasonably well with the ZLC studies of Brandani et al. [110] and with the frequency response studies of Song et al. [112], which both led to activation energies for the corrected surface diffusion coefficients of the xylene isomers in silicalite-1 in the temperature range of 100-400 °C within 20-30 kJ/mol. The studies of Wu et al. [113] and Xomeritakis et al. [109] showed that the activation energies of the adsorption constants of xylene isomers in ZSM-5 varied from 50 to 70 kJ/mol. The activation energies of K_i obtained here are 62±2, 51±3 and 58±6 kJ/mol for *p*-, *m*-, and *o*-xylene, respectively.

The agreement of these estimated diffusion and adsorption parameters with the data of previous researchers provides evidence that the model can be an alternative tool in the investigation of diffusion and adsorption in polycrystalline microporous media.

Table 7-4: Activation energies of D_i^S and K_i determined via fitting predicted single gas permeations of xylene isomers at 200 °C with experimental data of Haag et al [13], adopted from [53].

Activation energy	***p*-xylene**	***m*-xylene**	***o*-xylene**
$E_a(D_i^S)$ kJ/mol	28.8±1.6	27.6±2.8	13.5±5.5
$E_a(K_i)$ kJ/mol	62±2	51±3	58±6

7.2.2 Estimation of kinetic parameters of xylene isomerization

Haag et al. [13] have performed the xylene isomerization over a ZSM-5 membrane at a pressure of 1 bar and a temperature of 300°C. The general triangular scheme shown in Figure 7-2 was found by Haag et al. [13] to describe the reaction kinetics of xylene isomerization in the ZSM-5 membrane best. They expressed the reaction rates of xylene isomerization based on the partial pressures. As opposed to this, the reaction rates here are introduced based on coverage as the reactions occur on the surface:

p-xylene: $r_1 = k_2\theta_2 + k_6\theta_3 - (k_1 + k_5)\theta_1,$ Eq. 7—15

m-xylene: $r_2 = k_1\theta_1 + k_4\theta_3 - (k_2 + k_3)\theta_2,$ Eq. 7—16

o-xylene: $r_3 = k_5\theta_1 + k_3\theta_2 - (k_4 + k_6)\theta_3.$ Eq. 7—17

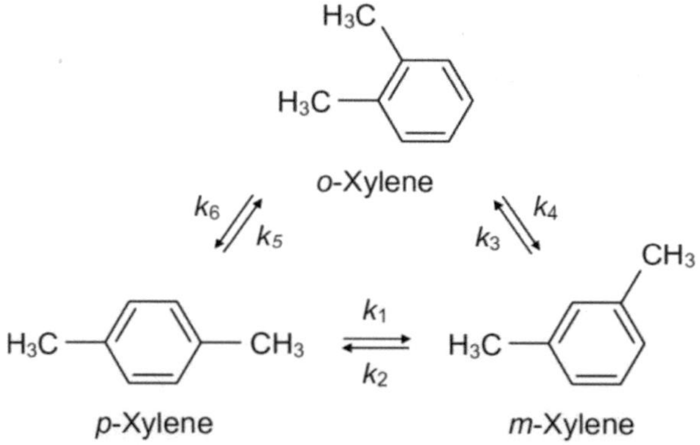

Figure 7-2. Triangular reaction scheme of xylene isomerization used by Haag et al. [13].

Thermodynamic equilibrium constants for xylene isomerization have been determined by Chirico and Steele [114] and are presented as follows:

$$K_{1,eq} = \frac{p_{o-xylene}}{p_{m-xylene}} = \exp(-791.8/T + 1575\times10^2/T^2 - 1417\times10^4/T^3)$$

Eq. 7—18

$$K_{2,eq} = \frac{p_{m-xylene}}{p_{p-xylene}} = \exp(0.693 + 4.792/T + 37070/T^2 - 5099\times10^3/T^3)$$

Eq. 7—19

$$K_{3,eq} = \frac{p_{p-xylene}}{p_{o-xylene}} = \exp(-0.693 + 787.1/T - 1946\times10^2/T^2 + 1927\times10^4/T^3)$$

Eq. 7—20

These expressions again are based on the partial pressures of the xylene isomers. Hence, they should be modified using the Langmuir equations before used in the CPNM. The partial pressures in Eq. 7—15 - Eq. 7—17 can be replaced by:

$$\theta_i = \frac{K_i p_i}{1+\sum_{j=1}^{n} K_j p_j} \Rightarrow p_i = \frac{\theta_i}{K_i} \frac{1}{1-\sum_{j=1}^{n}\theta_i}.$$

Eq. 7—21

So one obtains

$$K_{1,eq} = \frac{p_{o-xylene}}{p_{m-xylene}} = \left(\frac{\theta_3}{K_3}\frac{1}{1-\sum_{j=1}^{n}\theta_i}\right) \Bigg/ \left(\frac{\theta_2}{K_2}\frac{1}{1-\sum_{j=1}^{n}\theta_i}\right) = \frac{\theta_3}{\theta_2}\cdot\frac{K_2}{K_3}$$

Eq. 7—22

$$K_{2,eq} = \frac{p_{m-xylene}}{p_{p-xylene}} = \left(\frac{\theta_2}{K_2}\frac{1}{1-\sum_{j=1}^{n}\theta_i}\right) \Bigg/ \left(\frac{\theta_1}{K_1}\frac{1}{1-\sum_{j=1}^{n}\theta_i}\right) = \frac{\theta_2}{\theta_1}\cdot\frac{K_1}{K_2}$$

Eq. 7—23

$$K_{3,eq} = \frac{p_{p-xylene}}{p_{o-xylene}} = \left(\frac{\theta_1}{K_1} \frac{1}{1-\sum_{j=1}^{n}\theta_i} \right) \Bigg/ \left(\frac{\theta_3}{K_3} \frac{1}{1-\sum_{j=1}^{n}\theta_i} \right) = \frac{\theta_1}{\theta_3} \cdot \frac{K_3}{K_1}$$

Eq. 7—24

Finally, the thermodynamic equilibrium constants for xylene isomerization based on the coverages of the xylene isomers read as follows:

$$\frac{\theta_3}{\theta_2} = \frac{k_3}{k_4} = K_{1,eq} \cdot \frac{K_3}{K_2},$$ Eq. 7—25

$$\frac{\theta_2}{\theta_1} = \frac{k_1}{k_2} = K_{2,eq} \cdot \frac{K_2}{K_1},$$ Eq. 7—26

$$\frac{\theta_1}{\theta_3} = \frac{k_6}{k_5} = K_{3,eq} \cdot \frac{K_1}{K_3}.$$ Eq. 7—27

k_1, k_3, and k_5 were chosen as kinetic parameters for fitting, so the other kinetic parameters could be calculated from Eq. 7—25 - Eq. 7—27 by:

$$k_4 = k_3 \Bigg/ \left(K_{1,eq} \cdot \frac{K_3}{K_2} \right),$$ Eq. 7—28

$$k_2 = k_1 \Bigg/ \left(K_{2,eq} \cdot \frac{K_2}{K_1} \right),$$ Eq. 7—29

$$k_6 = \left(K_{3,eq} \cdot \frac{K_1}{K_3} \right) \cdot k_5 \cdot$$ Eq. 7—30

For estimation of the kinetic parameters of xylene isomerization, the adsorption and diffusion parameters previously obtained and shown in Table 7-2 - Table 7-4 were used. *Via* fitting to the experimental data of Haag et al.

[13] on xylene isomerization, the kinetic parameters at 300 °C were obtained and are shown in Table 7-5. The fitting results are given in Table A2 in Appendix d).

Table 7-5: Kinetics parameters determined *via* fitting with experimental data of the xylene isomerization [13] at 300 °C, adopted from [53].

i	**1**	**2**	**3**	**4**	**5**	**6**
$k_{i,573K}$ [mol kg^{-1}s^{-1}]	18.63	6.58	3.80	9.03	4.44	3.72

7.2.3 Investigation of the effects of structural parameters

Once the adsorption, diffusion and reaction parameters had been established, the effects of the structural parameters of the polycrystalline H-ZSM-5 layer on its performance in xylene isomerization were studied with the help of the simulation model. Two different cases were investigated: 1) the situation at low conversion corresponding to the experiments performed by Haag et al. [13], and 2) the situation at higher conversion, i.e., reached *via* a 100-fold increase of the contact time (membrane area). The observed trends were generally independent of the conversion level. An example is given for the influence of crystal orientation on selectivity and conversion in section 7.2.3.2.

7.2.3.1 Effects of concentration and connectivity of defects

As mentioned above, Knudsen diffusion is the dominant mechanism of transport in the defects (intercrystalline mesopores). The defects have no separation for the xylene isomers, since the isomers all have the same molecular molar mass. They can be isolated or connected with other defects forming an intercrystalline mesopore network. This defect pore network also connects to the intracrystalline pore system through the interface nodes, bringing about the possibility to access the micropores inside the crystallites *via* the intercrystalline mesopore network. Hence the reaction and separation performance of a polycrystalline layer will be affected not only by the number of defects per unit area, i.e., its concentration, but also by the connectivity of the defects. In the simulations, the concentration of the defects (i.e., the intercrystalline porosity ξ_{inter}) was varied in the range of 1×10^{-2} to 1×10^{-7}, whereas the connectivity of the defects was varied between 0 and 4.41.

Figure 7-3 gives an example of how the connectivity of the defect pores C influence selectivity and conversion in xylene isomerization when staring from *o*-xylene. The connectivity of the defect pores C as a function of the connection probability P of the defect pores ($C = 0.0441P$) has been shown in Figure 3-3. The error bars in Figure 7-3 indicate the variation of the *p*-xylene selectivity and *o*-xylene conversion among five consecutive calculations (standard errors) with the same value of P, i.e., the connection probability of the defect pores. It is shown that of the *p*-xylene selectivity is more or less independent of the connectivity of the defects whereas the *o*-xylene conversion increases with increasing connectivity of the intercrystalline mesopores.

Figure 7-4 shows the predicted selectivity towards *p*-xylene and conversion of *m*- or *o*-xylene at 300°C as a function of the intercrystalline porosity. The selectivity increases and the conversion decreases significantly with reduced intercrystalline porosity, when the intercrystalline porosity decreases from 1×10^{-2} to 1×10^{-4}. This is because more molecules will diffuse through the membrane *via* the intracrystalline micropores when decreasing the number of defects. Lower permeance of the xylene isomers is obtained but higher selectivity towards *p*-xylene. When the intercrystalline porosity falls below 1×10^{-4}, most of the xylene isomers diffuse through the membrane *via* the intracrystalline micropores. Thus, the selectivity towards *p*-xylene and the conversion of *m*- or *o*-xylene does no longer vary with decreasing concentration of the defects.

Consequently, the simulation results suggest a careful control of the concentration of defects by appropriate actions during membrane preparation or a post synthetic treatment to reduce it afterwards in order to improve the selectivity towards the product *p*-xylene.

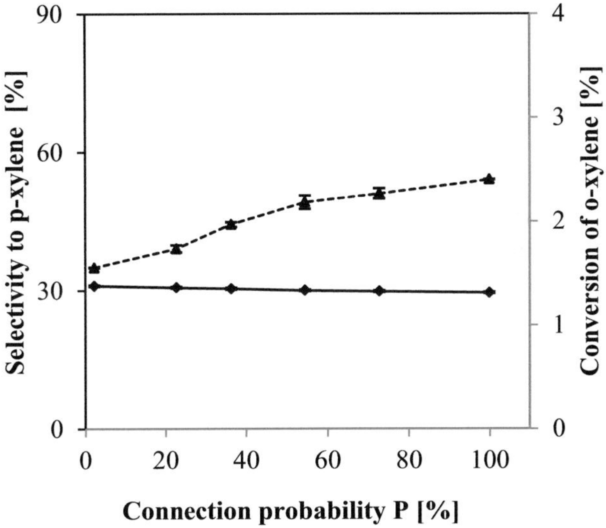

Figure 7-3: Dependence of *p*-xylene selectivity (solid line) and *o*-xylene conversion (dashed line) at 300 °C on the connection probability. Porosity of intercrystalline mesopores: 9×10^{-3}. Error bars give the standard errors for five consecutive calculations, adopted from [53].

a)

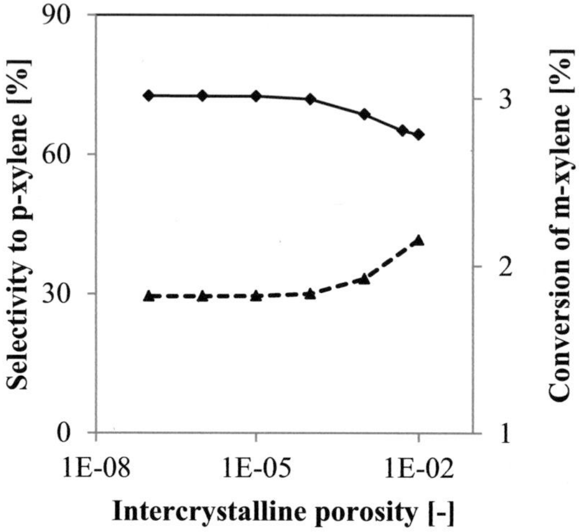

b)

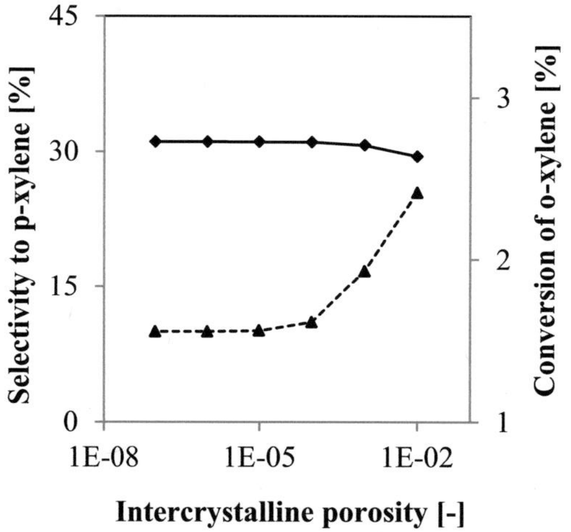

Figure 7-4: Dependence of p-xylene selectivity (solid line) and m- or o-xylene conversion (dashed line) at 300 °C on the intercrystalline porosity: a) m-xylene as Feed; b) o-xylene as Feed. Connection probability of intercrystalline mesopores P: 100 %, i.e. C = 4.41, adopted from [53].

7.2.3.2 Effect of crystallite orientation

Due to the anisotropic pore system, crystallite orientation is thought to play an important role on the catalytic performance of polycrystalline membranes, at least for high-quality almost defect-free layers, in which the reactants diffuse mainly through the intracrystalline micropores. In the investigation of the effects of crystallite orientation on xylene isomerization, the intercrystalline porosity was set to a low level of 1×10^{-7} to ensure that this condition is fulfilled. Random variation of the orientation angle of the crystallites was allowed for a more realistic representation of a polycrystalline layer. Figure 7-5 summarizes the effects of crystal orientation on selectivity and conversion for isomerization of *m*-xylene and *o*-xylene at low and high conversion levels for a reaction temperature of 300°C. In each case a normal distribution of the orientation angle Ψ (see Figure 7-1) with a standard deviation of 5° was assumed and five consecutive simulations were performed. The error bars indicate the randomness effect. Note that $\Psi = 0°$ means that the *b*-direction of the ZSM-5 crystallites is parallel to the membrane surface whereas for $\Psi = 90°$ it is perpendicular to the membrane surface.

Both for low and high conversion levels, the selectivity to *p*-xylene decreases and the conversion of *m*- and *o*-xylene increases with increasing angle Ψ. The increase of conversion is caused by faster reactant supply into the catalytic membrane at higher angle Ψ. For $\Psi = 90°$ the trans-membrane flux is maximized as the straight pores in the crystallites, which offer high diffusivity, run parallel to the membrane thickness. Hence, with increasing angle Ψ the concentration of xylene isomers inside the membrane rises. At low angle, high *p*-xylene selectivity is obtained, as the reaction is limited by reactant supply so that the loading remains below the limiting value where single file diffusion occurs. In this regime, *p*-xylene formed by the isomerization reaction quickly diffuses out of the membrane, giving rise to high *p*-xylene selectivity. At higher angle Ψ, the diffusive flux into the membrane is increased. In turn, the concentration of the xylene isomers in the membrane gets larger and single-file diffusion gains control. In this regime, the faster moving *p*-xylene is slowed down by the slower moving *o*- and *m*-xylene and is increasingly converted back to *m*-xylene. This is indicated in Table 7-5 showing that the constant k_1 describing the conversion of *p*-xylene to *m*-xylene is much larger than all other rate constants.

These results indicate that high selectivity to *p*-xylene can only be obtained when the loading of the xylene isomers in the membrane is low.

Additional calculations were performed to study whether a layer with randomly oriented crystallites (i.e., $\Psi = \text{Rand}(1)\times 90°$) can be represented by a layer with a uniform orientation angle of Ψ = 45 °. Five consecutive calculations were carried out. The predicted conversions and selectivities as well as their standard errors are summarized in Table 7-6. For a uniform orientation angle of 45° slightly higher conversions and lower *p*-xylene selectivity are detected than for random orientation of the crystallites. However, differences are very moderate, also when compared to the standard errors. Hence, a uniform orientation angle of 45° appears to be valid as a first approximation of a layer with random orientation of the crystallites.

7.2.3.3 Effect of crystallite size

Besides crystallite orientation, the effect of the crystallite size on the catalytic performance of the ZSM-5 membrane was investigated. The intercrystalline porosity and the angle Ψ were set to 1×10^{-7} and 45°, respectively. The crystallite size was varied between 0.75 and 3 μm while maintaining the same thickness of the polycrystalline layer, which means the number of intracrystalline nodes on top of each other was varied accordingly. Interfacial resistances between the intergrown crystallites, e.g., due to pore blockage or surface barrier effects were neglected at this stage.

Table 7-6: Predicted xylene conversions and *p*-xylene selectivities for two cases:
1) all crystallites aligned at $\Psi = 45°$ and
2) crystallites randomly orientated, i.e., $\Psi = \text{Rand}(1)\times 90°$, adopted from [53].

Feed	**Conversion (%)**		$S_{p\text{-xylene}}$ **%**		$S_{m\text{-xylene}}$ **%**		$S_{o\text{-xylene}}$ **%**	
Case	1)	2)	1)	2)	1)	2)	1)	2)
***m*-xylene**	1.82	1.67±0.09	72.5	73.3±1.1	-	-	27.5	26.7±1.1
	31.9	31.4±0.2	47.4	49.3±0.6	-	-	52.6	50.7±0.6
***o*-xylene**	1.55	1.40±0.12	31.1	31.4±0.4	68.9	68.6±0.4	-	-
	51.3	49.4±1.3	23.2	23.6±0.1	76.8	76.4±0.1	-	-

a)

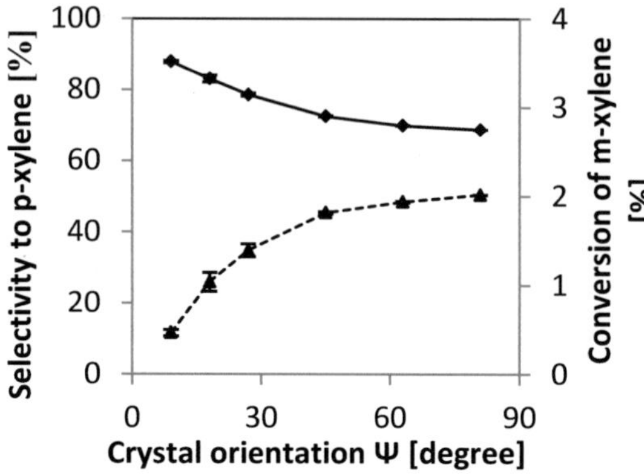
100
80
60
40
20
0
Selectivity to p-xylene [%]
4
3
2
1
0
Conversion of m-xylene [%]
0
30
60
90
Crystal orientation Ψ [degree]

b)

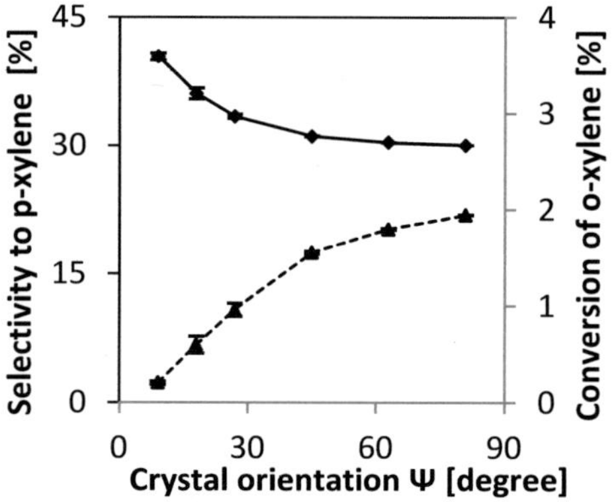
45
30
15
0
Selectivity to p-xylene [%]
4
3
2
1
0
Conversion of o-xylene [%]
0
30
60
90
Crystal orientation Ψ [degree]

c)

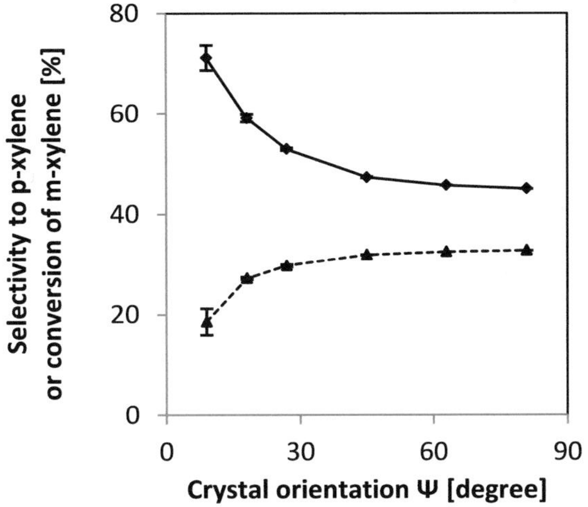

d)

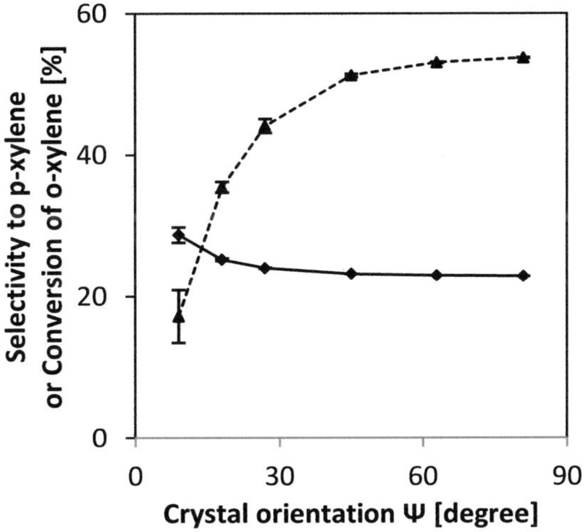

Figure 7-5: Dependence of selectivities to *p*-xylene (solid line) and conversions of reactants (dashed line) at 300 °C on the angle Ψ describing the crystallite orientation: a) *m*-xylene as Feed, low conversions; b) *o*-xylene as Feed, low conversions; c) *m*-xylene as Feed, high conversions; d) *o*-xylene as Feed, high conversions, adopted from [53].

Figure 7-6 shows the selectivity to *p*-xylene and the conversions of *m*- and *o*-xylene at 300 °C as a function of the crystallite size. Selectivity to *p*-xylene decreases slightly and conversion increases with decreasing crystallite size. This can be explained by the fact that the concentration inside the crystallites is taken as uniform in the present model. Therefore, like in a cascade of stirred tank reactors, conversion is increased with increasing number of units because on average the concentration becomes larger. Due to the reaction kinetics of xylene isomerization, the selectivity to *p*-xylene shows the opposite trend.

In reality, however, for bulky molecules like xylene isomers interfacial effects are expected to become more important for the whole intracrystalline transport when the crystallites are smaller. In addition to pore blockage the interfacial effects also include a barrier of entrance into the intracrystalline pores and the existence of a physisorbed precursor state on the outside of the zeolite crystallite. For example, the experimental results of Zheng et al. [111] show that the diffusion coefficients of xylene isomers in smaller ZSM-5 crystallites are lower than those in large crystallites, in particular for *p*-xylene. This implies that the interfacial effects are more significant for smaller crystallites. For the calculations shown in Figure 7-6, interfacial effects were ignored because no quantified information on the magnitude of the interfacial resistance is available. However, taking these effects into account could well lead to a reversed trend, i.e., higher conversion for larger crystals, because then fewer interfaces between the crystallites exist. To investigate this more in detail experimental data from polycrystalline layers with distinctly different size of the crystallites would be required as well as a more accurate model taking into account the interfacial effects.

a)

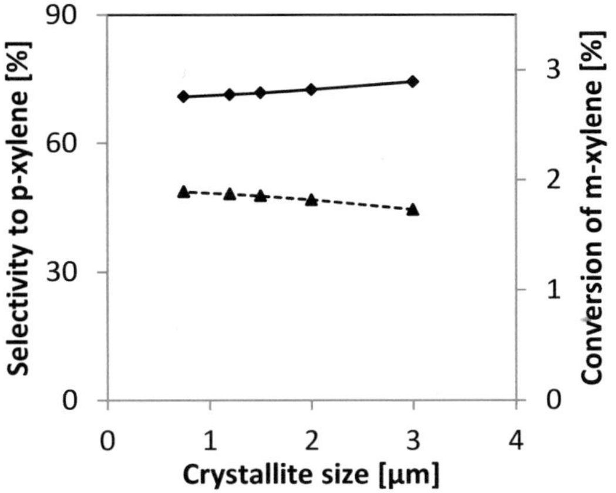

b)

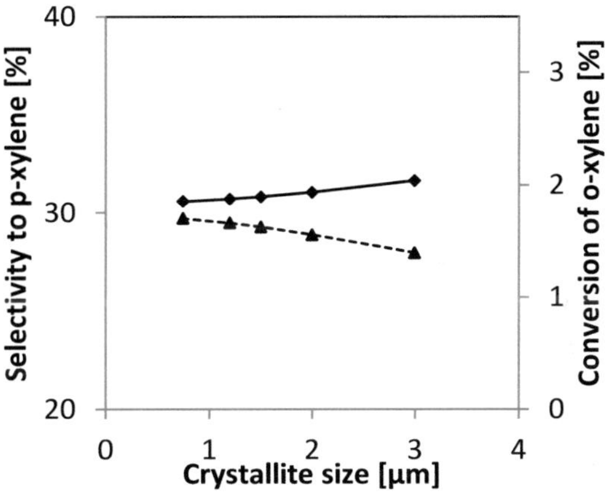

Figure 7-6: Dependence of *p*-xylene selectivity (solid line) and conversion (dashed line) at 300 °C on the crystallite size: a) *m*-xylene as Feed, low conversion; b) *o*-xylene as Feed, low conversion, adopted from [53].

7.3 Conclusions

Effects of some structural parameters, such as the concentration and connectivity of defects, the crystallite orientation and size, were investigated for xylene isomerization in a catalytic H-ZSM-5 membrane. Simulation results show that the intercrystalline pores and the crystallite orientation in H-ZSM-5 membranes affect the catalytic performance on xylene isomerization significantly. The crystallite size has little effect. Therefore, it is suggested that the concentration and connectivity of the defects should be controlled by appropriate methods in the preparation of H-ZSM-5 membranes at a low level or reduced after membrane preparation, in order to achieve a high selectivity towards *p*-xylene. Moreover, the simulation shows that increasing the amount of the *b*-oriented crystallites can increase the concentration of the xylene isomers in the crystallites and therefore the conversion of *m*- and *o*-xylene to *p*-xylene, but at the cost of reduced selectivity towards *p*-xylene, since the effect of single-file diffusion is significant at high concentration of the xylene isomers in the crystallites.

8 Summary and outlook

The research in this dissertation focused on the development of a simulation-assisted design method based on experiments and simulations to accelerate the development of hierarchically structured catalyst systems with polycrystalline and anisotropic zeolite layers. This simulation-assisted design method has been successfully applied to two examples: bifunctional core-shell catalysts with a H-ZSM-5 shell layer ($Cu/ZnO/Al_2O_3$@H-ZSM-5) for direct DME synthesis from synthesis gas, and catalytic H-ZSM-5 membranes for xylene isomerization.

The main achievement of the modeling and simulation work is the development of a new 3-D pore network model – "crystallite-pore network model", which can simulate the polycrystalline and anisotropic structures of the zeolite catalysts. In this model, diffusion models like SDM, SFDM and KDM were used to simulate the diffusion in the pore space of the zeolite catalysts. The model is applicable to many practical problems since reactions represented by any type of kinetic expression are allowed. Using this model, the structural effect of the H-ZSM-5 shell layer in the core-shell catalyst, as well as the catalytic H-ZSM-5 membranes for xylene isomerization has been studied, e.g., the concentration of defect pores, the crystallite orientation, the shell thickness. Moreover, some guidances on the catalyst preparation and optimization were also given from the simulation results. However, the calculation of the CPNM is very time-consuming, in particular when the network size of the CPNM is large. Therefore, parallel computing technology and a multicore super computer are required to reduce the calculation time of the CPNM, in order to make it more applicable to practical problems, e.g., for determination of reaction parameters in the core-shell catalyst *via* regression method.

In this dissertation, besides the CPNM, a heterogeneous model under the assumption of a pseudo-homogeneous shell was used to simulate diffusion and reaction within the core-shell catalyst particles. For determination of the reaction parameters in the core-shell catalyst, a reactor model based on this heterogeneous model was developed for a tubular reactor, which was packed with the prepared core-shell catalyst particles to obtain the catalyst test data. This much less time-consuming simulation model has been successfully used to determine the reaction parameters of the prepared core-shell catalyst particles using the experimental data from the tubular reactor.

On the experimental side, a high-temperature high-pressure adsorption measurement device was devised based on a high frequency LCM oscillating microbalance together with colleagues at IMVT. It has been successfully used to analyze the adsorption of carbon dioxide, water, methanol and dimethyl ether in the H-SZM-5 (Si/Al = 100) zeolite in the temperature range of 50-150 °C and pressure range of 0-18 bar. The adsorption isotherms obtained from the adsorption measurements using the LCM adsorption measurement device were fitted with the single and dual site Langmuir models. The results show that the adsorption of CO_2, methanol and DME in H-ZSM-5 are well described by the single site Langmuir model, while the dual site Langmuir model is needed for fitting the adsorption of H_2O in H-ZSM-5. Moreover, the adsorption capacities, enthalpies and entropies for these gases derived from the Langmuir models were generally in good agreement with literature data. Under the reaction conditions of direct DME synthesis, the adsorption capacities of these gases in H-ZSM-5 obey the following order: CO_2 > H_2O (site 1) > methanol > DME > H_2O (site 2), while the adsorption enthalpies are ordered as: H_2O (site 1) > H_2O (site 2) > methanol > DME > CO_2. In the forthcoming work, the LCM device will be used to investigate the adsorption behavior of gas mixtures present in the direct DME synthesis reaction. Furthermore, the ideal absorbed solution theory (IAST) proposed by Myers and Prausnitz [45] will be used to simulate the adsorption behavior of gas mixtures in H-ZSM-5 using the single component data determined in this study. The experimental data of gas mixtures from the LCM device can be used to validate the IAST.

As another achievement in experiments, a high-pressure and high-temperature WK cell was established to measure the permeation of gases through the H-ZSM-5 membrane for determination of the corrected surface diffusion coefficients of gases in H-ZSM-5, which are required in simulation of diffusion in H-ZSM-5. The test results *via* permeation experiments of a gas mixture of CO_2 and H_2 show that this apparatus has a good tightness even at high temperatures and pressures. Moreover, the sealing on the zeolite membrane is also quite good. Thus, this apparatus and membrane is ready to determine the size and concentration of defects in the prepared H-ZSM-5 membrane *via* e.g., *n*-hexane permporosimetry measurements, and afterwards the corrected surface diffusion coefficients of gases in the H-ZSM-5 zeolite *via* single permeation measurements. In addition, this apparatus can be used as a membrane reactor for methanol dehydration to validate the CPNM with the measured gas compisotions in the retentate and permeate chambers. A single-pellet diffusion reactor, which had a similar structure as the WK cell, has been successfully used by Rieckmann & Keil [115] to validate their 3-D regular random pore network models for porous catalyst pellets.

Future extension of the 3-D crystallite-pore network model will focus on overcoming the limitation of the current model that concentration gradients inside the crystallites cannot be simulated. A possible approach is using a secondary cubic pore network with several nodes (e.g. 5×5×5) to represent every crystallite. An improvement on the modeling of transport in micropores is the application of the IAST [45] on the modeling of the adsorption of gas mixtures, in order to simulate the coverages of components more accurately, which are important for the SDM to simulate the multicomponent diffusion in the micropores. Moreover, the model for diffusion in micropores can be improved by considering the interfacial effects [116], particularly when the activation energy for entrance in the pores and exit from the pores is larger than the activation energy for intracrystalline diffusion. Bonilla and Bhatia [117] developed an oscillator model from statistical-mechanical principles, which showed that it could predict diffusivities in nanopores more accurately than the Knudsen diffusion model. In the future work, the oscillator model may be considered to improve the modeling of transport in the defects of the polycrystalline media, instead of the present idea Knudsen diffusion model.

References

[1] J. K. Nørskov, T. Bligaard, J. Rossmeisl, C. H. christensen, Towards the Computational Design of Solid Catalysts, *Nature Chem.*, **2009**, 1, 37-46.

[2] B. Hammer, J. K. Nørskov, Theoretical Surface Science and Catalysis - Calculations and Concepts, *Adv. Catal.*, **2000**, 45, 71-129.

[3] X. Du, X. Gao, W. Hu, J. Yu, Z. Luo, K. Cen, Catalyst Design Based on DFT Calculations: Metal Oxide Catalysts for Gas Phase NO Rreactantion, *J. Phys. Chem. C*, **2014**, 118(25), 13617-13622.

[4] M. P. Andersson, T. Bligaard, A. Kustov, K. E. Larsen, J. Greeley, T. Johannessen, C. H. Christensen, J. K. Nørskov, Toward Computational Screening in Heterogeneous Catalysis: Pareto-Optimal Methanation Catalysts, *J. Catal.*, **2006**, 239, 501-506.

[5] L. C. Grabow, M. Mavrikakis, Mechanism of Methanol Synthesis on Cu through CO_2 and CO Hydrogenation, *ACS Catal.*, **2011**, 1, 365-384.

[6] J. Cejka, A. Corma, S. Zones, Zeolites and Catalysis: Synthesis, Reactions and Applications, John Wiley & Sons, **2010**.

[7] X. Zhu, S. Liu, Y. Song, L. Xu, Catalytic Cracking of C4 Alkenes to Propene and Ethene: Influences of Zeolites Pore Structures and Si/Al_2 Tatios, *Applied Catalysis A: General*, **2005**, 288, 134-142.

[8] J. Bao, J. He, Y. Zhang, Y. Yoneyama, N. Tsubaki, A Core/Shell Catalyst Produces a Spatially Confined Effect and Shape Selectivity in a Consecutive Reaction, *Angew. Chem.*, **2007**, 120, 359-362.

[9] G. Yang, N. Tsubaki, J. Shamoto, Y. Yoneyama, Y. Zhang, Confinement Effect and Synergistic Function of H-ZSM-5/Cu-ZnO-Al_2O_3 Capsule Catalyst for One-Step Controlled Synthesis, *J. Am. Chem. Soc.*, **2010**, 132, 8129-8136.

[10] G. Yang, M. Thongkam, T. Vitidsant, Y. Yoneyama, Y. Tan, N. Tsubaki, A Double-Shell Capsule Catalyst with Core-Shell-like Structure for One-Step exactly Controlled Synthesis of Dimethyl Ether from CO_2 Containing Syngas, *Catal. Today*, **2011**, 229-235.

[11] A. M. Tarditi, S. Irusta, E. A. Lombardo, Xylene Isomerization in a Membrane Reactor: Part I: The Synthesis of MFI Membranes for the *p*-Xylene Separation, *Chem. Eng. J.*, **2006**, 122, 167-174.

[12] A. L. Deshayes, E. E. Miro, G. I. Horowitz, Xylene Isomerization in a Membrane Reactor Part II. Simulation of an Industrial Reactor, *Chem. Eng. J.*, **2006**, 122, 149-157.

[13] S. Haag, M. Hanebuth, G. T. P. Mabande, A. Avhale, W. Schwieger, R. Dittmeyer, On the Use of a Catalytic H-ZSM-5 Membrane for Xylene Isomerization, *Micropor. Mesopor. Mater.*, **2006**, 96, 168-176.

[14] U. Hong, J. Kärger, R. Kramer, H. Pfeifer, G. Seiffert, PFG n.m.r. Study of Diffusion anisotropy in oriented ZSM-5 type zeolite crystallites, *Zeolites*, **1991**, 11, 816-821.

[15] J. Kärger, Random Walk through Two-Channel Networks: A Simple Means To Correlate the Coefficients of Anisotropic Diffusion in ZSM-5 Type Zeolites, *J. Phys. Chem.*, **1991**, 95, 5558-5560.

[16] Z. Lai, G. Bonilla, I. Diaz, J. G. Nery, K. Sujaoti, M. A. Amat, E. Kokkoli, O. Terasaki, R. W. Thompson, M. Tsapatsis, D. G. Vlachos, Microstructural Optimization of a Zeolite Membrane for Organic Vapor Separation, *Science*, **2003**, 300, 456-460.

[17] A. O. Koriabkina, PhD Dissertation: Diffusion of Alkanes in MFI-type Zeolites, Eindhoven University of Technology, The Netherlands, **2003**.

[18] R. Krishna, Multicomponent Surface Diffusion of Adsorbed Species: a Description Based on the Generalized Maxwell-Stefan Equations, *Chem. Eng. Sci.*, **1990**, 45, 1779-1791.

[19] A. Wheeler, Reaction Rates and Selectivity in Catalyst Pores, *Adv. Catal.*, **1951**, 3, 249-327.

[20] S. Reyes, E. Iglesia, K. F. Jensen, Application of Percolation Theory Concepts to the Analysis of Gas-Solid Reactions, *Solid State Ionics*, **1989**, 32-33(2), 833-842.

[21] M. P. Hollewand, L. F. Gladden, Modeling of Diffusion and Reaction in Porous Catalysts using a Random Three-Dimensional Network Model, *Chem. Eng. Sci.*, **1992**, 47(7), 1761-1770.

[22] N. A. Seaton, Determination of the Connectivity of Porous Solids from Nitrogen Sorption Measurements, *Chem. Eng. Sci.*, **1991**, 46(8), 1895-1909.

[23] J. Wood, L. F. Gladden, Modeling Diffusion and Reaction Accompanied by Capillary Condensation Using Three-Dimensional Pore Networks. Part 1. Fickian Diffusion and Pseudo-First-Order Reaction Kinetics, *Chem. Eng. Sci.*, **2002a**, 57, 3033-3045.

[24] C. Rieckmann, F. J. Keil, Multicomponent Diffusion and Reaction in Three-Dimensional Networks: General Kinetics, *Ind. Eng. Chem. Res.*, **1997**, 36, 3275-3281.

[25] J. Wood, L. F. Gladden, F. J. Keil, Modeling Diffusion and Reaction Accompanied by Capillary Condensation Using Three-Dimensional Pore Networks. Part 2. Dusty Gas Model and General Reaction Kinetics, *Chem. Eng. Sci.*, **2002b**, 57, 3047-3059.

[26] M. Dadvar, M. Sahimi, Pore Network Model of Deactivation of Immobilized Glucose Isomerase in Packed-Bed Reactors. Part III: Multiscale Modeling, *Chem. Eng. Sci.*, **2003**, 58, 493-4951.

[27] F. Chen, R. Mourhatch, T. T. Tsotsis, M. Sahimi, Pore Network Model of Transport and Separation of Binary Gas Mixtures in Microporous Membranes, *J. Membr. Sci.*, **2008**, 315, 48-57.

[28] G. M. Laudone, G. P. Matthews, P. A. C. Gane, Modeling Diffusion from Simulated Porous Structures, *Chem. Eng. Sci.*, **2008**, 63, 1987-1996.

[29] A. Johnson, I. M. Roy, G. P. Matthews, D. Patel, An Improved Simulation of Void Structure Water Retention and Hydraulic Conductivity in Soil, Using the Pore-Cor Three-Dimensional Network. *Eur. J. Soil Sci.*, **2003**, 54, 477-489.

[30] P. Bodurtha, G. P. Matthews, J. P. Kettle, I. M. Roy, Influence of Anisotropy on the Dynamic Wetting and Permeation of Paper Coatings. *J. Colloid Interface Sci.*, **2005**, 283, 171-189.

[31] J. T. Gostick, Random Pore Network Modeling of GDLs Using Voronoi and Delaunay Tessellations, *ECS Transactions*, **2011**, 41(1), 125-130.

[32] K. E. Thompson, Pore-Scale Modeling of Fluid Transport in Disordered Fibrous Materials, *AIChE J.*, **2002**, 48(7), 1369-1389.

[33] E. A. Mason, A. P. Malinauskas, Gas Transport in Porous Media: the Dusty-Gas Model, *Chemical engineering monographs* 17, Elsevier, Amsterdam, **1983**.

[34] J. Xiao, J. Wei, Diffusion Mechanism of Hydrocarbons in Zeolites - I. Theory. *Chem. Eng. Sci.*, **1992a**, 47, 1123-1141.

[35] F. J. Keil, Diffusion and Reaction in Porous Networks, *Catal. Today*, **1999**, 53, 245-258.

[36] E. N. Fuller, P. D. Schettler, J. C. Giddings, New Method for Prediction of Binary Gas-Phase Diffusion Coefficients, *Ind. Eng. Chem.*, **1966**, 58(5), 18-27.

[37] J. R. Welty, C. E. Wicks, R. E. Wilson, G. L. Rorrer, *Fundamentals of Momentum, Heat and Mass Transfer*, 5th edition. Hoboken: John Wiley and Sons, **2008**.

[38] R. Mezaki, H. Inoue, Rate Equations of Solid Catalysed Reactions. Tokyo: University of Tokyo Press, **1991**.

[39] J. M. Van de Graaf, F. Kapteijn, J. A. Moulijn, Modeling Permeation of Binary Mixtures through Zeolite Membranes, *AIChE J.*, **1999**, 45, 497-511.

[40] D. Paschek, R. Krishna, Inter-Relation between Self- and Jump-Diffusivities in Zeolites, *Chem. Phys. Lett.*, **2001**, 333, 278-284.

[41] U. Hong, J. Kärger, R. Kramer, H. Pfeifer, G. Seiffert, PFG n.m.r. Study of Diffusion Anisotropy in Oriented ZSM-5 Type Zeolite Crystallites, *Zeolites*, **1991**, 11, 816-821.

[42] H. Jobic, M. Bée, J. Caro, M. Bülow, J. Kärger, Molecular Self-Diffusion of Methane in Zeolite ZSM-5 by Quasi-Elastic Neutron Scattering and Nuclear Magnetic Resonance Pulsed Field Gradient Technique, *J. Chem. Soc., Faraday Trans. 1*, **1989**, 85, 4201-4209.

[43] P. Demontis, E. S. Fois, G. B. Suffritti, S. Quartieri, Molecular Dynamics Studies on Zeolites. 4. Diffusion of Methane in Silicalite, *J. Phys. Chem.*, **1990**, 94(10), 4329-4334.

[44] F. Kapteijn, J. A. Moulijn, R. Krishna, The Generalized Maxwell-Stefan Model for Diffusion in Zeolites: Sorbate Molecules with Different Saturation Loadings, *Chem. Eng. Sci.*, **2000**, 55, 2923-2930.

[45] A. L. Myers, J. M. Prausnitz, Thermodynamics of Mixed-Gas Adsorption, *AICHE J.*, **1965**, 11, 121-127.

[46] J. Kärger, D. M. Ruthven, D. N. Theodorou, *Diffusion in Nanoporous Materials*, Wiley-VCH Verlag GmbH & Co. KGaA, Weinheim, Germany, **2012**, Chapter 5.

[47] C. D. Baertsch, H. H. Funke, J. L. Falconer, R. D. Noble, Permeation of Aromatic Hydrocarbon Vapors through Silicalite–Zeolite Membranes, *J. Phys. Chem.*, **1996**, 100(18), 7676–7679.

[48] G. Xomeritakis, Z. Lai, M. Tsapatsis, Separation of Xylene Isomer Vapors with Oriented MFI Membranes Made by Seeded Growth. *Ind. Eng. Chem. Res.*, **2001**, 40, 544-552.

[49] L. J. P. Van den Broeke, W. J. W. Bakker, F. Kapteijn, J. A. Moulijn, Binary Permeation through a Silicalite-1 Membrane, *AIChE J.*, **1999**, 45, 976-985.

[50] IUPAC Manual of Symbols and Terminology for Physicochemical Quantities and Units, Butterworths, London, **1972**.

[51] F. J. Keil, *Modeling Reactions in Porous Media*, in *Modeling and Simulation of Heterogeneous Catalytic Reactions: From the Molecular Process to the Technical System* (ed O. Deutschmann), Wiley-VCH Verlag GmbH & Co. KGaA, Weinheim, Germany, **2011**.

[52] Y. Yan, M .E. Davis, G. R. Gavalas, Preparation of Highly Selective Zeolite ZSM-5 Membranes by a Post-Synthetic Coking Treatment. J. *Membr. Sci.*, **1997**, 123(1), 95-103.

[53] W. Ding, H. Li, P. Pfeifer, R. Dittmeyer, Crystallite-Pore Network Model of Transport and Reaction of Multicomponent Gas Mixtures in Polycrystalline Microporous Media, *Chem. Eng. J.*, **2014**, 254, 545-558.

[54] M. C. Lovallo, A. Gouzinis, M. Tsapatsis, Synthesis and Characterization of Oriented MFI Membranes Prepared by Secondary Growth. *AIChE J.*, **1998**, 44, 1903-1913.

[55] M. J. D. Powell, A Fortran Subroutine for Solving Systems of Nonlinear Algebraic Equations, in Numerical Methods for Nonlinear Algebraic Equations, P. Rabinowitz ed., London, Gordon and Breach, 115-161, **1970**.

[56] J. U. Keller, R. Staudt, *Gas Adsorption Equilibria: Experimental Methods and Adsorptive Isotherms*. Springer Science & Business Media: New York, USA, **2005**.

[57] Y. Belmabkhout, M. Frere, G. De Weireld, High-Pressure Adsorption Measurements. A Comparative Study of the Volumetric and Gravimetric Methods. *Meas. Sci. Technol.*, **2004**, 15, 848-858.

[58] T. C. Golden, S. Sircar, Gas Adsorption on Silicalite. *J. Colloid Interface Sci.*, **1994**, 162, 182.

[59] M. S. Sun, D. B. Shah, H. H. Xu, O. Talu, Adsorption Equilibria of C_1 to C_4 Alkanes, CO_2, and SF_6 on Silicalite. *J. Phys. Chem. B*, **1998**, 102, 1466-1473.

[60] V. Tsionsky, E. Gileadi, Use of the Quartz Crystal Microbalance for the Study of Adsorption from the Gas Phase. *Langmuir*, **1994**, 10, 2830-2835.

[61] A. Venkatasubramanian, M. Navaei, K. R. Bagnall, K.C. McCarley, S. Nair, P. J. Hesketh, Gas Adsorption Characteristics of Metal–Organic Frameworks *via* Quartz Crystal Microbalance Techniques. *J. Phys. Chem. C*, **2012**, 116, 15313-15321.

[62] G. Sauerbrey, Verwendung von Schwingquartzen zur Wägung dünner Schichten und Zur Mikrowägung. *Zeitschrift für Physik*, **1959**, 155, 206-222.

[63] P. M. Davulis, M. Pereira da Cunha, High-Temperature Langatate Elastic Constants and Experimental Validation up to 900 °C. *IEEE Transactions on Ultrasonics, Ferroelectrics, and Frequency Control*, **2010**, 57, 59-65.

[64] W. Ding, M. Klumpp, H. Li, U. Schygulla, P. Pfeifer, W. Schwieger, K. Haas-Santo, R. Dittmeyer, Investigation of High-Temperature and High-Pressure Gas Adsorption in Zeolite H-ZSM-5 *via* the Langatate Crystal Microbalance: CO_2, H_2O, Methanol, and Dimethyl Ether, *J. Phys. Chem. C*, **2015**, 119(41), 23478-23485.

[65] W. Ding, G. Baracchini, M. Klumpp, W. Schwieger, R. Dittmeyer, Adsorption Device Based on a Langatate Crystal Microbalance for High Temperature High Pressure Gas Adsorption in Zeolite H-ZSM-5, *J. Vis. Exp.*, **2016** (submitted).
[66] O. de la Iglesia, V. Sebastián, R. Mallada, G. Nikolaidis, J. Coronas, G. Kolb, R. Zapf, V. Hessel, J. Santamaría, Preparation of Pt/ZSM-5 Films on Stainless Steel Microreactors. *Catal. Today*, **2007**, 125, 2-10.
[67] S. K. Wirawan; D. Creaser, CO_2 Adsorption on Silicalite-1 and Caution Exchanged ZSM-5 Zeolites Using a Step Change Response Method, *Microporous Mesoporous Mater.*, **2006**, 91, 196–205.
[68] V. R. Choudhary; S. Mayadevi, Adsorption of Methane, Ethane, Ethylene, and Carbon Dioxide on High Silica Pentasil Zeolites and Zeolite-like Materials Using Gas Chromatography Pulse Technique, *Sep. Sci. Technol.*, **1993**, 28, 2197-2209.
[69] V. R. Choudhary, S. Mayadevi, Adsorption of Methane, Ethane, Ethylene, and Carbon Dioxide on Silicalite-I, *Zeolites*, **1996**, 17, 501-507.
[70] W. D. Zhu; P. Hrabanek, L. Gora, F. Kapteijn, J.A. Moulijn, Role of Adsorption in the Permeation of CH_4 and CO_2 through a Silicalite-1 Membrane, *Ind. Eng. Chem. Res.*, **2006**, 45, 767-776.
[71] D. H. Olson, W. O. Haag, W.S. Borghard, Use of Water as a Probe of Zeolite Properties: Interaction of Water with HZSM-5, *Microporous Mesoporous Mater.*, **2000**, 35-36, 435-446.
[72] V. Bolis, C. Busco, Thermodynamic Study of Water Adsorption in High-Silica Zeolites, *J. Phys. Chem. B*, **2006**, 110, 14849-14859.
[73] L. Ohlin, P. Bazin, F. Thibault-Starzyk, J. Hedlund, M. Grahn, Adsorption of CO_2, CH_4, and H_2O in Zeolite ZSM-5 Studied Using in situ ATR-FTIR Spectroscopy, *J. Phys. Chem. C*, **2013**, 117(33), 1697-16982.
[74] C. Pope, Water Adsorption on ZSM-5 and its Aluminum Free Analog, Silicalite, *J. Colloid Interface Sci.*, **1987**, 116(1), 221-223.
[75] M. Hunger, D. Freude, H. Pfeifer, Magic-Angle Spinning Nuclear Magnetic Resonance Studies of Water Molecules Adsorbed on

Brønsted- and Lewis-acid Sites in Zeolites and Amorphous Silica-Aluminas, *J. Chem. Soc. Faraday Trans.*, **1991**, 87(4), 657-662.

[76] A. Jentys, G. Warecka, M. Derewinski, J. A. Lercher, Adsorption of Water on ZSM5 Zeolites, *J. Phys. Chem.*, **1989**, 93(12), 4837-4843.

[77] B. Hanger, S. Matysik, M. Heuchel, W. Einicke, Adsorption of Methanol on ZSM-5 Zeolites, *Langmuir*, **1997**, 13, 6249-6254.

[78] V. S. Nayak, J. B. Moffat, Sorption and Diffusion of Alcohols in Heteropoly Oxometalates and ZSM-5 Zeolite, *J. Phys. Chem.*, **1988**, 92, 7097-7102.

[79] C. Pope, Adsorption of Methanol and Related Molecules on Zeolite H-ZSM-5 and Silicalite, *J. Chem. Soc. Faraday Trans.*, **1993**, 89(7), 1139-1141.

[80] D. Ambrose, J. H. Ellender, C. H. S. Sprake, R. Townsend, Thermodynamic Properties of Organic Oxygen Compounds XLIII. Vapour Pressures of Some Ethers, *J. Chem. Thermodynamics*, **1976**, 8(2), 165-178.

[81] A. Avhale, Ph.D. dissertation: *Development of Stainless-Steel Supported MFI and BEA Type Zeolite Membranes*, Friedrich-Alexander-University of Erlangen-Nürnberg, Erlangen, **2010.**

[82] M. Yu, J. L. Falconer, and R. D. Noble, Characterizing Nonzeolite Pores in MFI Membranes, *Ind. Eng. Chem. Res.*, **2008**, 47, 3943-3948.

[83] S. Tauro, PhD dissertation: *One-Step Synthesis of Dimethyl Ether Using Microreactors*, Karlsruhe Institute of Technology, Shaker Verlag, Aachen, **2014**.

[84] W. Ding, M. Klumpp, S. Lee, S. Reuß, S. A. Al-Thabaiti, P. Pfeifer, W. Schwieger, R. Dittmeyer, Simulation of One-Stage Dimethyl Ether Synthesis over Core/Shell Catalyst in Tube Reactor, *Chemie Ingenieur Technik*, **2015**, 87, 702-712.

[85] *Review of evidence on health aspects of air pollution – REVIHAAP project: final technical report.* WHO/Europe, **2013**. http://www.euro.who.int/__data/assets/pdf_file/0004/193108/REVIHAAP-Final-technical-report-final-version.pdf?ua=1.

[86] B. Brunekreef, S. T. Holgate, Air Pollution and Health, *The Lancet*, **2002**, 360, 1233-1242.

[87] E. D. Larson, H. Yang, Dimethyl Ether (DME) from Coal as a Household Cooking Fuel in China, *Energy for Sustainable Development*, **2004**, 3(3), 115-126.
[88] T. A. Semelsberger, R. L. Borup, H. L. Greene, Dimethyl Ether (DME) as an Alternative Fuel, *Journal of Power Sources*, **2006**, 156, 497-511.
[89] Z. Azizi, M. Rezaeimanesh, T. Tohidian, M. R. Rahimpour, Dimethyl Ether: A Review of Technologies and Production Challenges, *Chem. Eng. Process.*, **2014**, 82, 150-172.
[90] Y. Wang, W. Wang, Y. Chen, J. Ma, R. Li, Synthesis of Dimethyl Ether from Syngas over Core-Shell Structure Catalyst Cu-ZnO-Al_2O_3@SiO_2-Al_2O_3, *Chem. Eng. J.*, **2014**, 250, 248-256.
[91] Z. Graaf, E. Stamhuis, A. Beenackers, Kinetics of Low Pressure Methanol Synthesis, *Chem. Eng. Sci.*, **1988**, 43, 3185-3195.
[92] G. Bercic, J. Levec, Catalytic Dehydration of Methanol to Dimethyl Ether. Kinetic Investigation and Reactor Simulation, *Ind. Eng. Chem. Res.*, **1993**, 32, 2478-2484.
[93] J. B. Anderson, A Criterion for Isothermal Behaviour of a Catalyst Pellet, *Chem. Eng. Sci.*, **1963**, 18, 147-148.
[94] D.E. Mears, Tests for Transport Limitations in Experimental Catalytic Reactors; *Ind. Eng. Chem. Process Des. Dev.*, **1971**, 10(4), 541-547.
[95] M. Baerns, H. Hoffmann, A. Renken, *Chemische Reaktionstechnik*, Thieme Georg, Stuttgart **1999**.
[96] J. Kärger, H. Pfeifer, F. Stallmach, ^{129}Xe and ^{13}C PFG n.m.r. Study of the Intracrystalline Self-Diffusion of Xe, CO2, and CO, *Zeolites*, **1993**, 13, 50-55.
[97] J. Caro, S. Hocevar, J. Kärger, L. Riekert, Intracrystalline Self-Diffusion of H_2O and CH_4 in ZSM-5 Zeolites; *Zeolites*, **1986**, 6, 213-216.
[98] S.K. Wirawan, D. Creaser, CO_2 Adsorption on Silicalite-1 and Cation Exchanged ZSM-5 Zeolites Using a Step Change Response Method, *Microporous Mesoporous Mater.*, **2006**, 91, 196-205.
[99] X. Du, E. Wu, Physisorption of Hydrogen in A, X and ZSM-5 Types of Zeolites at Moderately High Pressures, *Chin. J. Chem. Phys.*, **2006**, 19(5), 457-462.

[100] X. Du, Y. Huang, E. Wu, Molecular-Dynamics Studies of the Diffusion of H_2 in All-Silica ZSM-5; *Materials Science Forum*, **2012**, 704-705, 401-406.

[101] A. Ison, R.J. Gorte, The Adsorption of Methanol and Water on H-ZSM-5, *J. Catal.*, **1984**, 89,150-158.

[102] S.K. Wirawan, D. Creaser, Multicomponent $H_2/CO/CO_2$ Adsorption on BaZSM-5 Zeolite, *Sep. and Pur. Tech.*, **2006**, 52, 224-231.

[103] J.A. Dunne, M. Rao, S. Sircar, R.J. Corte. A.L. Myers, Calorimetric Heats of Adsorption and Adsorption Isotherms. 2. O_2, N_2, Ar, CO_2, CH_4, C_2H_6, and SF_6 on NaX, H-ZSM-5, and Na-ZSM-5 Zeolites, *Langmuir*, **1996**, 12, 5896-5904.

[104] G.H. Graaf, P.J.J.M. Sijtsema, E.J. Stamhuis, G.E.H. Joosten, Chemical Equilibria in Methanol Synthesis, *Chem. Eng. Sci.*, **1986**, 41, 2883-2890.

[105] W. Lu, L. Teng, W. Xiao, Simulation and Experiment Study of Dimethyl Ether Synthesis from Syngas in a Fluidized-Bed Reactor, *Chem. Eng. Sci.*, **2004**, 59, 5455-5464.

[106] M.J.D. Powell, *Numerical Methods for Nonlinear Algebraic Equations*, P. Rabinowitz, ed., Ch.7, London, **1970**.

[107] D. H. Olson, G. T. Kokotailo, S. L. Lawton, W. M. Meier, Crystal Structure and Structure-Related Properties of ZSM-5, *J. Phys. Chem.*, **1981**, 85(15), 2238-2243.

[108] X. Du, E. Wu, Porosity of Microporous Zeolites A, X and ZSM-5 Studied by Small Angle X-ray Scattering and Nitrogen Adsorption, *J. Phys. Chem. Solids*, **2007**, 68, 1692-1699.

[109] G. Xomeritakis, S. Nair, M. Tsapatsis, Transport Properties of Alumina-Supported MFI Membranes Made by Secondary (Seeded) Growth, *Micropor. Mesopor. Mater.*, **2000**, 38, 61-73.

[110] S. Brandani, M. Jama, D.M. Ruthven, Counter-Diffusion of *p*-Xylene/Benzene and *p*-Xylene/*o*-Xylene in Silicalite Studied by the Zero-Length Column Technique, *Ind. Eng. Chem. Res.*, **2000**, 39, 821-828.

[111] S. Zheng, A. Jentys, J.A. Lercher, Xylene Isomerization with Surface-Modified HZSM-5 Zeolite Catalysts: An in situ IR study, *J. Catal.*, **2006**, 241, 304-311.

[112] L. Song, L.V.C. Rees, Adsorption and Diffusion of Cyclic Hydrocarbon in MFI-Type Zeolites Studied by Gravimetric and Frequency-Response Techniques, *Micropor. Mesopor. Mater.*, **2000**, 35-36, 301-314.

[113] P. Wu, A. Debebe, Y. Ma, Adsorption and Diffusion of C6 and C8 Hydrocarbons in Silicalite, *Zeolites*, **1983**, 3, 118-122.

[114] R. Chirico, W.V. Steele, Thermodynamic Equilibria in Xylene Isomerization. 5. Xylene Isomerization Equilibria from Thermodynamic Studies and Reconciliation of Calculated and Experimental Product Distributions. *J. Chem. Eng. Data*, **1997**, 42, 784-790.

[115] C. Rieckmann, F. J. Keil, Simulation and Experiment of Multicomponent Diffusion and Reaction in Three-Dimensional Networks, *Chem. Eng. Sci.*, **1999**, 54, 3485-3493

[116] R. M. Barrer, Porous Crystal Membranes, *J. Chem. Soc. Faraday Trans.*, **1990**, 86(7), 1123-1130.

[117] M. R. Bonilla, S. K. Bhatia, The Low-Density Diffusion Coefficient of Soft-Sphere Fluids in Nanopores: Accurate Correlations from Exact Theory and Criteria for Applicability of the Knudsen Model, J. Membr. Sci., 2011, 382, 339–349.

Nomenclature

A =	area, m^2
b =	proportion of blocking pores, -
c =	concentration, mol m^{-3}
$\mathbf{c}$ =	vector of concentration
C =	connectivity of intercrystalline pores, -
d =	diameter, m
D =	diffusion coefficient, m^2 s^{-1}
$\mathbf{D}$ =	diffusive matrix
E_a =	activation energy, J mol^{-1}
J =	flux, mol m^{-2} s^{-1}
J' =	flux after scaling, mol m^{-2} s^{-1}
k =	reaction rate, m^3 kg^{-1} s^{-1}
K =	adsorption constant, Pa^{-1}
l =	length of intercrystalline pores, µm
L =	reactor length, m
m =	mass, kg
M =	molar mass, kg mol^{-1}

n =	number of components, number of discretized layers or sections, -
N =	flow, mol s^{-1}
N' =	flux after scaling, mol s^{-1}
p =	pressure, bar
P =	connection possibility of intercrystalline pores, %
q =	loading, mol kg^{-1}
$\mathbf{r}$ =	vector of reaction rate, mol m^{-2} s^{-1}
$\boldsymbol{R}$ =	gas constant, J mol^{-1} K^{-1}
S =	selectivity to products, or membrane selectivity, -
T =	temperature, K
V =	volume, m^{-3}
X =	conversion, -
y =	molar fraction, -

Greek letters

α =	factor of flux distribution in x-, y- and z-coordinate, -
β =	mass transfer coefficient, m s^{-1}
ε =	porosity, -
θ =	coverage, -

μ = chemical potential, J mol^{-1}

$\mathbf{v}$ = matrix of stoichiometric numbers, -

ρ = density, kg m^{-3}

ς = scaling of pore number, -

ψ = angle between the *b*-direction of crystal and *xy*-flat, degree

χ = angle of the projection line of *b*-direction on the *xy*-flat with *x*-coordinate, degree

τ = tortuosity, -

$\mathbf{\Gamma}$ = thermodynamic matrix

Subscripts

bulk = bulk phase

cat = catalysts

cry = crystallite

eq = equilibrium

ex = experimental

feed = feed gas

in = inlet

inter = intercrystalline

intra = intracrystalline

ip = serial number of pores connected with the node i

i, j = component number

K = Knudsen diffusion

memb = membrane

meso = mesopore

net = network

outside = outside surface of the network, out of one reaction section

p = pore, particle

sat = saturation

self = self-diffusion

sim = simulated

sweep = sweep gas

tot = total

Superscripts

feed = feed gas

ip = pores connected at the node representing a crystallite

per = permeate side

ret = retentate side

sweep =	sweep gas
S =	surface diffusion

Appendix

a) Adsorption device based on LCM

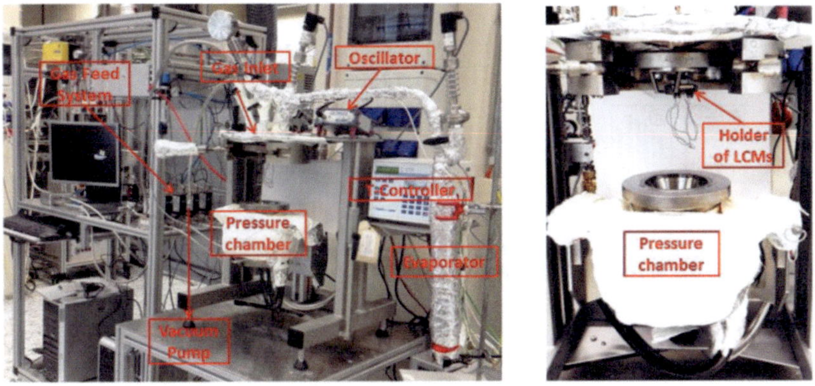

Figure A1: Overview of the LCM-based adsorption measurement apparatus (left), and detailed view of the high temperature, high pressure stainless steel chamber with the holder of LCM chips (right), adopted from the Supporting Information of [64].

Table A1: Detailed experimental data and calculated values for adsorption of CO_2 in H-ZSM-5 at 50 °C from the LCM device, adopted from the Supporting Information of [64].

p bar	$f_{0,r}$ Hz	Δf_0 Hz	f_s Hz	Δf Hz	Δm mg	q mmol/g
15.50±0.05	14881070±30	2350	14862550±50	16170±80	0.57531±0.00285	3.327±0.129
14.14±0.05	14880950±30	2350	14862420±30	16180±60	0.57568±0.00213	3.344±0.097
11.04±0.05	14880750±20	2350	14862300±30	16100±50	0.57285±0.00178	3.215±0.081
9.65±0.05	14880500±10	2350	14862150±50	16000±60	0.56931±0.00213	3.055±0.097
7.10±0.05	14880190±10	2350	14862000±50	15840±60	0.56364±0.00213	2.798±0.097
3.78±0.05	14879040±10	2350	14861220±50	15470±60	0.55056±0.00213	2.206±0.097
2.29±0.05	14878570±10	2350	14861050±50	15170±60	0.53991±0.00213	1.724±0.097
1.08±0.05	14879300±10	2350	14862150±50	14800±60	0.52669±0.00213	1.125±0.097
0.28±0.03	14878020±10	2350	14861220±50	14450±60	0.51433±0.00213	0.565±0.097
0.00±0.03	14878370±10	2350	14861920±50	14100±60	0.50185±0.00213	0.000±0.097

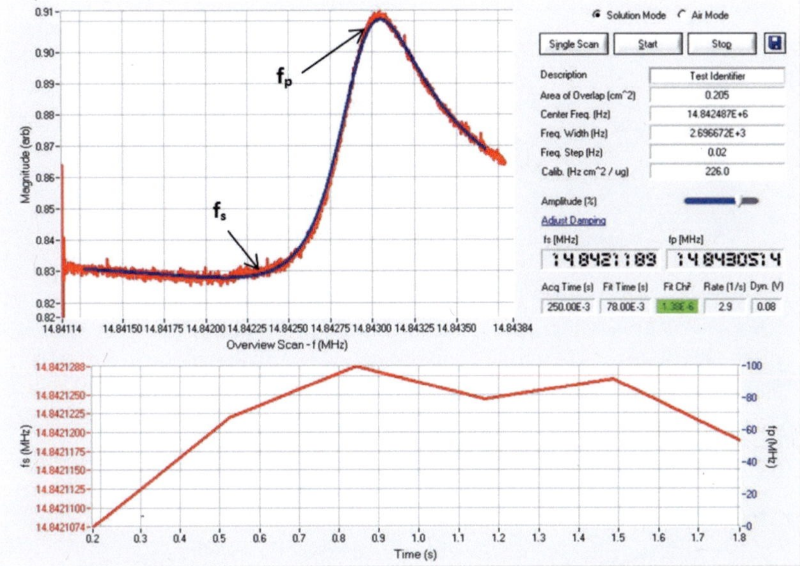

Figure A2: Resonant frequency measurement of LCM loaded with H-SZM-5 (150 °C, 0.3 bar methanol) by the Gamry's Resonator™ software *via* fitting with Butterworth – van Dyke Equivalent Circuit Model. Upper diagram: fit diagram; bottom diagram: f_s vs. time. f_s and f_p represent the series resonant frequency and the parallel resonant frequency, respectively. In this study, f_s was chosen to calculate the adsorption data, adopted from the Supporting Information of [64].

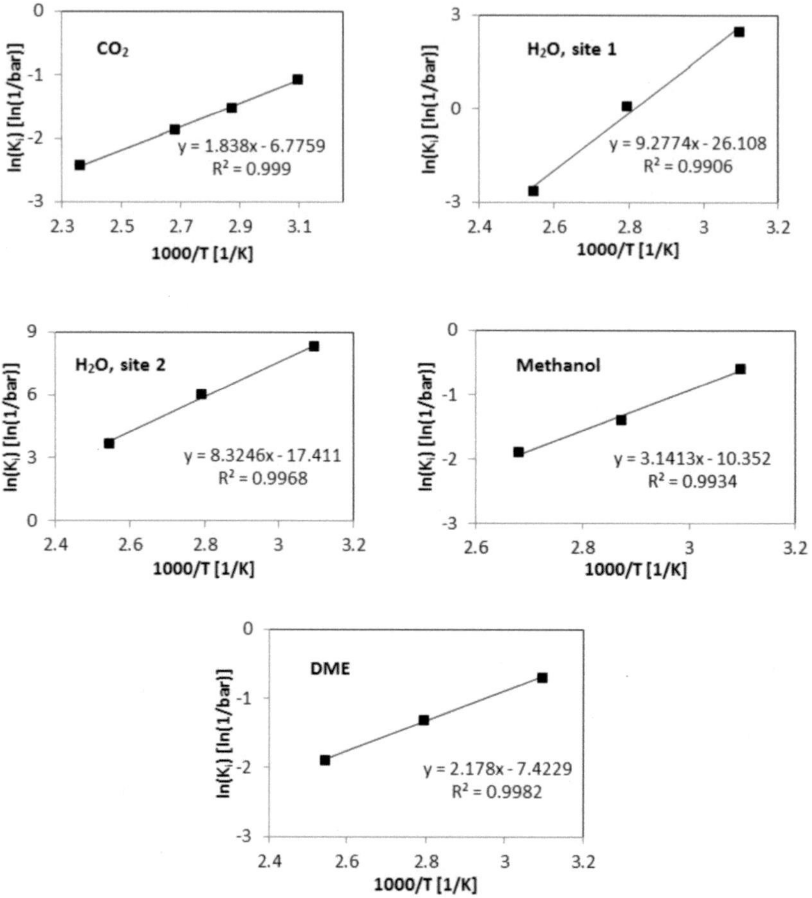

Figure A3: ln(K_i) vs. 1000/T to determine adsorption enthalpies and entropies for CO_2, water (site 1 and site 2), methanol and DME, adopted from the Supporting Information of [64].

b) High temperature high-pressure Wicke-Kallenbach cell

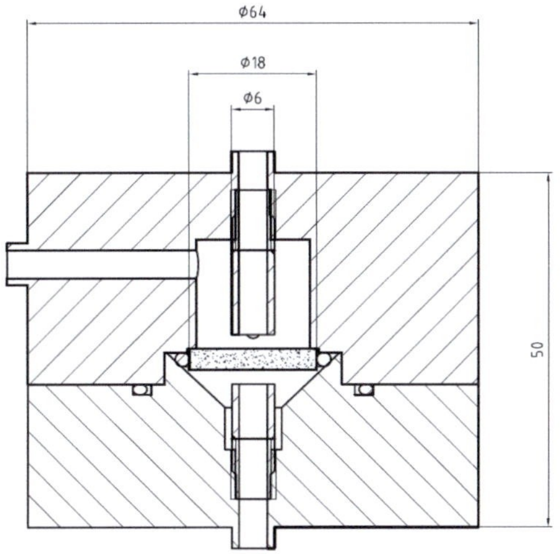

Figure A4: Dimensions of Wicke-Kallenbach cell fabricated in IMVT, with the unit of mm.

Figure A5: Photograph of high-temperature high-pressure Wicke-Kallenbach cell fabricated in IMVT for permeation test on membranes

c) Berty-type reactor

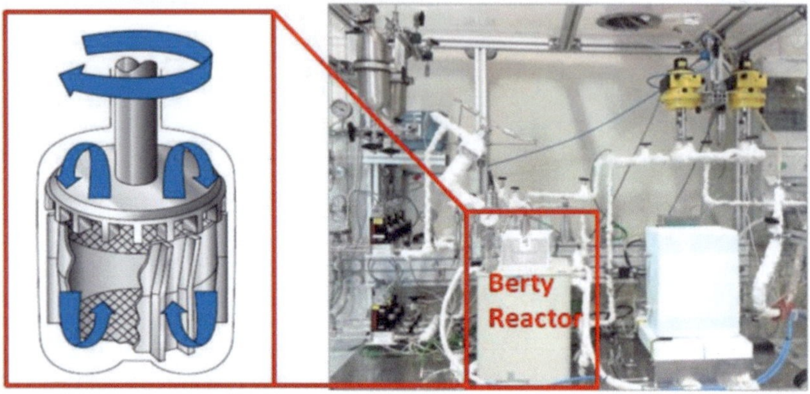

Figure A6: A concentration gradient-free Berty-type reactor for catalyst test.

d) CNPM for xylene isomerization

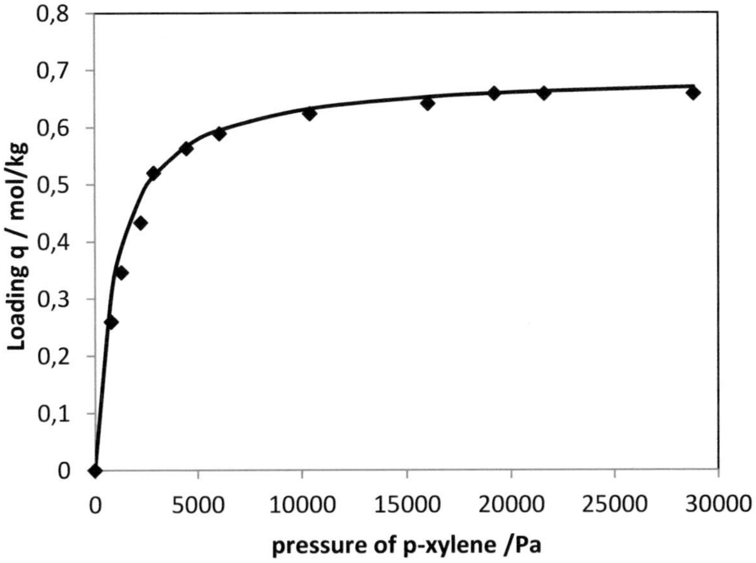

Figure A7: Single-component adsorption isotherm of p-xylene in a ZSM-5 membrane at 100 °C. Symbols: experimental data of Xomeritakis et al. [109]; Line: Fitted data with the Langmuir equation, adopted from [53].

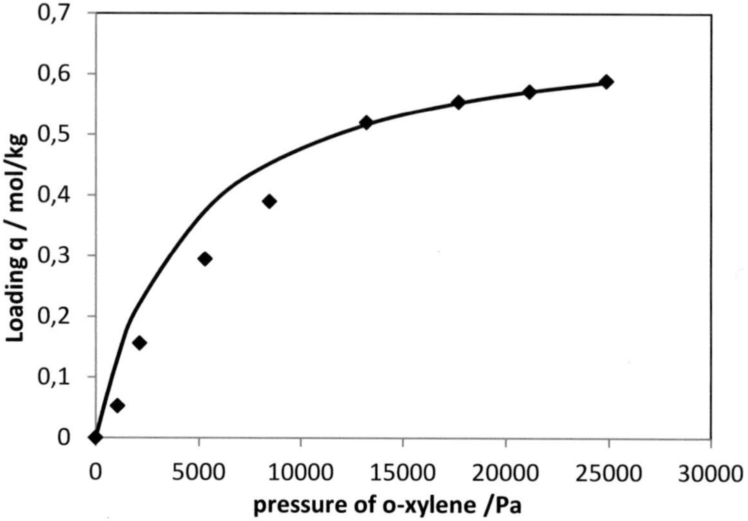

Figure A8: Single-component adsorption isotherms of o-xylene in a ZSM-5 membrane at 100 °C. Symbols: experimental data of Xomeritakis et al. [109]; Line: Fitted data with the Langmuir equation, adopted from [53].

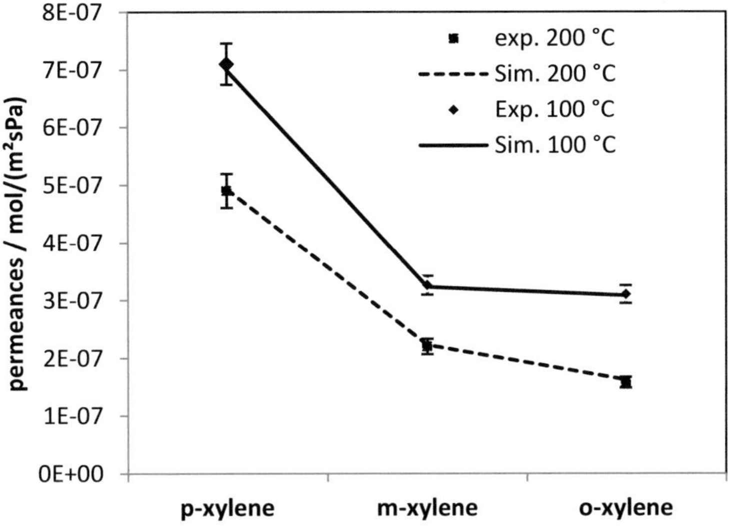

Figure A9: Predicted and experimentally determined single gas permeances of xylene isomers at 100 and 200 °C. The experimental permeances have the measure errors of 6 %, adopted from [53].

Table A2: Comparison of predicted values with experimental data of the xylene isomerization in the ZSM-5 membrane reactor at 300 °C. Error = (Sim. Value – Exp. Value)/ Exp. Value×100%, adopted from [53].

Feed	**Conversion (%)**			$S_{p\text{-}xylene}$ %			$S_{m\text{-}xylene}$ %			$S_{o\text{-}xylene}$ %		
	Exp.	Sim.	Error	Exp.	Sim.	Error	Exp.	Sim.	Error	Exp.	Sim.	Error
***p*-xylene**	5.6	5.9	+5%	-	-	-	82.6	81.2	-2%	17.4	18.8	+8%
***m*-xylene**	2.2	2.0	-9%	63.6	64.4	+1%	-	-	-	36.4	35.6	-2%
***o*-xylene**	2.6	2.5	-4%	30.8	29.5	-4%	69.2	70.5	+2%	-	-	-

Lists of Figures

Lists of Tables